SKILLS for EXAM SUCCESS MATHS

Joelene Kearney

Gill Education
Hume Avenue
Park West
Dublin 12
www.gilleducation.ie

Gill Education is an imprint of M.H. Gill & Co.

© Joelene Kearney 2024

ISBN: 978-0-7171-99761

All rights reserved. No part of this publication may be copied, reproduced or transmitted in any form or by any means without written permission of the publishers or else under the terms of any licence permitting limited copying issued by the Irish Copyright Licensing Agency.

Design: emc design ltd

Illustrations: Elaine Leggett, D'Avila Illustration

At the time of going to press, all web addresses were active and contained information relevant to the topics in this book. Gill Education does not, however, accept responsibility for the content or views contained on these websites. Content, views and addresses may change beyond the publisher or author's control. Students should always be supervised when reviewing websites.

For permission to reproduce photographs, the authors and publisher gratefully acknowledge the following:
© Adobe Stock: 7, 13, 17, 18, 20, 22, 24, 26, 29, 30, 32, 43, 49, 53, 67, 68, 140, 152, 188, 189, 191, 204, 209, 217; © iStock/Getty Premium: 226, 243.

The author and publisher have made every effort to trace all copyright holders. If, however, any have been inadvertently overlooked, we would be pleased to make the necessary arrangement at the first opportunity.

Acknowledgements

This book is dedicated to Mam and Dad – thank you for everything you do for us every day.

A special and sincere word of thanks to the following for all your help, support, guidance and, above all, patience, dedication and hard work in getting this book to print: Emily Lynch, Fiona Watson, Anne Sophie Blytmann, Emma O'Donoghue, Durga Vemula, Eric Pradel, Hannah McAdams and the design team at emc design ltd. Thank you all so much.

Joelene Kearney

Table of Contents

	Learning Outcomes	Page
Introduction		iv
Chapter 1: Numbers	U1, U2, U3, U4, U5, U6, U7, U8, U11, U13, N1, N2	1
Chapter 2: Applied Arithmetic	U1, U2, U3, U4, U5, U6, U7, U8, U9, U11, U13, N1, N2, N3, GT4	17
Chapter 3: Sets	U1, U2, U3, U4, U5, U6, U7, U8, U9, U11, U13, N1, N2, N3, N5, SP2, SP3, AF4	34
Chapter 4: Algebra 1	U1, U2, U4, U5, U6, U7, U8, U11, U13, N2, AF2, AF3, GT2	48
Chapter 5: Algebra 2	U1, U2, U3, U4, U5, U6, U7, U8, U9, U10, U11, U13, N1, N4, AF1, AF2, AF3, AF4, AF6	60
Chapter 6: Algebra 3	U1, U2, U3, U4, U5, U6, U7, U8, U9, U10, U11, U13, N1, N2, N4, GT2, GT4, AF1, AF3, AF4, AF5, AF7	77
Chapter 7: Co-ordinate Geometry of the Line	U1, U2, U3, U4, U5, U6, U7, U8, U9, U11, U13, GT2, GT5, GT6, AF2, AF4	93
Chapter 8: Number Patterns	U1, U2, U3, U4, U5, U6, U7, U8, U9, U10, U11, U13, N1, N3, N4, GT1, GT2, GT3, GT5, AF1, AF2, AF4, AF7	109
Chapter 9: Functions and Graphing Functions	U1, U2, U3, U4, U5, U6, U7, U8, U9, U10, U11, U13, N1, N3, NF4, GT5, AF1, AF2, AF4, AF5, AF7	124
Chapter 10: Trigonometry	U1, U2, U3, U4, U5, U6, U7, U8, U9, U10, U11, U13, N1, N3, N4, GT3, GT4, GT5, GT6, AF1, AF2, AF3, AF4, AF7	144
Chapter 11: Geometry 1	U1, U2, U3, U5, U6, U7, U8, U10, U11, U12, U13, N2, GT3, GT4, AF2, AF4	161
Chapter 12: Applied Measure	U1, U2, U3, U5, U6, U7, U8, U9, U13, N1, N2, GT1, GT2, GT3, GT4, AF1, AF2	178
Chapter 13: Statistics 1	U1, U2, U3, U4, U5, U6, U7, U8, U9, U11, U13, N1, AF2, AF3, SP1, SP2, SP3	197
Chapter 14: Statistics 2	U1, U2, U3, U4, U5, U6, U7, U8, U9, U11, U13, N1, N2, N3, AF2, AF3, GT5, SP1, SP2, SP3	212
Chapter 15: Probability	U1, U2, U3, U4, U5, U6, U7, U8, U9, U11, U13, N1, N5, GT4, GT5, AF4, SP1, SP2, SP3	225
Chapter 16: Geometry 2	U1, U2, U3, U4, U5, U6, U7, U8, U9, U11, U12, U13, N1, N3, GT3, GT4, GT5, AF2, AF3, AF4	241
Chapter 17: Geometry 3	U1, U2, U5, U6, U7, U8, U9, U11, U12, U13, N3, GT1, GT2, GT3, GT4, GT5, GT6, AF2, AF4	259
Answers		273

Introduction

'But how do you study maths?' 'It's impossible to study maths.' If I had a euro for every time these two statements have been directed at me over the years – well, let's just say the villa in Spain would be a reality and not a dream.

And the truth of it is, it really is hard to 'study' maths. In my opinion, you can only truly understand maths and evaluate what you can actually do or not do by working through the problems, running into difficulties and figuring out which of your skills you can use to get to the solution. Simply reading a question or worked example and saying 'Yeah, I get that' does not mean you actually do. In my opinion, sitting down with a question that you have never seen before and working through it really and truly is the only way to study and prepare for your maths exam. The reality is that your maths exam is going to be exactly that – sitting down for two hours with a set of questions you have never seen before and using the skills that you have learnt over the last three years (and the eight years before that in primary school) to find the solution. The more practice you have at that beforehand, the more comfortable and confident you will be going into your Junior Cycle exam.

My favourite feature of this book is that if you do get stumped by any of the questions, there is a tutorial for you to watch, showing you how to work through the required steps to find the solution – a handy little hint if you get stuck. The tutorials are the most important feature for me. When I was in school, we had the answers at the back of the book so we could check if we were right or wrong. But I found it really frustrating that if I got stuck with a question, I could find out what the answer was but there was nothing to show me how to get it! Here, if you do get stuck, you simply scan the QR code and there's a step-by-step worked solution, clearly explaining each stage.

It is my hope that this book will help you through your preparation journey and will facilitate independent learning and study – skills you will be able to use throughout the rest of your education. It is a book filled with questions from every strand of the syllabus which have been designed to challenge you and sharpen your mathematical skills, with a back-up plan of tutorials explaining every step in every question if you need a little push in the right direction. I really hope it helps!

The very best of luck to you all!

Joelene

Advice for Exam Success

The Prep

- The night before the exam is not for learning new material. The work is done at this stage. The information is in your brain. You have developed the skills necessary to answer any question. Spend a couple of hours going over your notes and exam papers and practising some questions to refresh the information.

- Get your pencil case organised and into your bag. Pack your maths set and make sure all the different components are actually in the case – protractor, set square, ruler, compass. Sharpen your pencils and ensure that you have plenty of biros, a rubber and a topper.

- Make sure your calculator is working and that you are familiar with the different settings you may have to use in your exam. Put it into your pencil case and into your bag.

- Cramming does not work. Please don't stay up all hours trying to force the information into your already tired and overwhelmed brain. Put the books away at a reasonable time, go for a walk or a run, watch TV, whatever helps you to unwind, and go to bed early. A good night's sleep is vital to success.

The Exam

- Get to the exam centre in plenty of time. Rushing will do nothing for your stress levels. Make sure you know what time the exam starts, and arrive on time.

- You will have two hours to complete your exam. The number of questions on the exam paper can change from year to year and you need to **answer all questions**.

- I recommend that you spend five minutes at the start of the exam reading through the full paper and highlighting the key words in the questions. As you are reading through the paper, your brain will be logging the various topics that you will be examined on. It is processing the information and different 'memory boxes of information' will be opening.

- Please do not panic if you see something that you don't recognise initially or understand. Focus on **what you can do and start with that!** Your brain will be working away in the background on the parts of the exam that you are not so confident with. You will be very pleasantly surprised by what you will remember and what inspiration you will have when you return to the problematic question.

- The questions are of varying lengths and the SEC gives a time guide for each question. Do keep an eye on the time. Don't get bogged down by a tricky question. Move on and come back to it at the end.

- Read the questions carefully. Highlight the key words and make sure you understand what you are being asked. If you are asked to give your answer in a particular format or asked to round your answer, highlight this too. If you do not give your answer in the required format, marks will be deducted.

- If you are working through a solution and you decide 'No, this is wrong', just draw a single line through the work so that it is still legible to the examiner. Correctors have to assign marks to every piece of work. If we can still read the work that you have crossed out, we have to correct it and assign marks. If it is in fact worth more marks than any subsequent attempts, these are the final marks that will be assigned to this part of the question. Because of this, I would advise not using correcting fluid or rubbing anything out.

- Answer the question within the space provided. If you need to use additional space, it is a nice gesture to the examiner to write at the end of the question 'please see page ___ for additional work'. The examiner can then go to that page to complete their correcting of the question.

- Make your work clear and easy to read for the examiner. If you are asked to draw a graph or a chart, use all the space provided – don't try to squash it into a corner of the answer box. Remember, your script is one of several hundred the examiner has to correct and the easier it is for them to read, the easier it is for you to get your marks.

- Watch out for two-part questions. For example, 'Are the triangles congruent? Justify your answer.' Two answers are required here and there are marks for both.

- If you are asked to give, for example, two advantages/disadvantages or two reasons, write each point on a separate line to make it as clear to the examiner as possible.

- If you are asked to mark something on a diagram, do so neatly and accurately and in a colour different to the one used in the diagram. This will make the corrector's job easier.

- When carrying out your calculations, **show all your work**. There are marks available for different steps in the process and you do not want to lose out. If you just write down the answer with no workings, you will not get full marks.

- Partial credit will be awarded for work of merit (WOM). The marking scheme is not designed until after the exam is done. This means none of us know exactly what will be classified as WOM until the marking scheme is finalised. My advice is to write down everything you know – write the formula from the log tables into your answer booklet; draw a diagram to create a visual of what you are being asked in the question; if a diagram is given, then mark on the diagram what you are being asked to find; if it's a geometry question, write down some facts you know about the shape being referenced.

- Even if you can only do part of a question or the first few steps, **do them!** Some marks are better than no marks and you may be pleasantly surprised with what you can do once you get started. **Please write something! Do not leave a blank space!** The examiner cannot give partial credit for a blank space and if your efforts are deemed WOM, you will get marks.

- If you get the answer to part (a) of a question wrong, you will lose marks. However, if you use the incorrect answer that you got in part (a) correctly in subsequent parts of the question, and finish these parts correctly, you will get full marks for those questions. You have been penalised in part (a), but you will not be penalised a second time within that question.

- This point cannot be emphasised enough – **please do not be thrown by a topic/question/graph/diagram/phrasing that you do not recognise. The questions will all be based on information that you already have.**

Topic-specific Advice

- Always be aware of your signs when adding/subtracting, multiplying/dividing.

- UNITS, UNITS, UNITS! Do not forget the units in your answer! The general rule of thumb is if you are asked to, for example, give your answer in m², then there are no marks assigned for units in your answer. However, if units are not mentioned in the question, marks are assigned for the units in your answer.

- If you are completing an applied measure question and are asked, for example, to give your answer correct to 2 decimal places but are not given the value for π, then use the π button on your calculator. The answer given in the marking scheme will be rounding according to this value for π.

- If you are asked to give your answer in a particular format – for example, 'Find the equation of a line and give your answer in the form $ax + by + c = 0$' – and you fail to give your answer in this format, then marks will be lost. However, do not let the fact that you may not be able to give your answer in the required format stop you from attempting the question and doing as much of it as you can.
- Similarly, if you are asked to give your answer to the nearest whole number/2 decimal places/3 significant figures and you fail to complete this last step in the question, marks will be lost.
- Specifically, if you are asked to prove if a point is/is not on a given line in co-ordinate geometry, it is important that you include a sentence explaining what your answer means. It is not sufficient to place a tick or a cross beside your equation, you have to write either 'one side of the equals is the same as the other, therefore the point is on the line' or 'one side of the equals does not equal the other, therefore the point is not on the line'.
- When carrying out a constructions in geometry, show all construction lines clearly.
- In trigonometry, it is not sufficient to write SOH CAH TOA in the answer box. To be in with a chance of gaining partial credit, you have to write out the trigonometric ratios in full:
  $\sin \theta = \dfrac{\text{opposite}}{\text{hypotenuse}}$
  $\cos \theta = \dfrac{\text{adjacent}}{\text{hypotenuse}}$
  $\tan \theta = \dfrac{\text{opposite}}{\text{adjacent}}$
- Always write the formula from the log tables into the answer booklet. As I've said, you don't know what may be considered WOM in the scheme.

- When graphing functions:
  1. Always use a pencil.
  2. Label the x- and the y-axes.
  3. Make sure the scales along the x- and the y-axes are constant.
  4. Plot the x co-ordinate first and then the y. For example, for the point $(-2, -7)$, go to -2 on the x-axis first, then -7 on the y-axis. Plot the points clearly and accurately.
  5. Join your points together. There are marks for plotting **and** for joining the points.
- When drawing a bar chart:
  1. Use a pencil.
  2. Label the axes.
  3. Each bar in the bar chart must be the same width and be clearly labelled.
  4. The **space** between the bars must be equal.
- When drawing a pie-chart:
  1. In the exam, there are marks for calculating the number of degrees dedicated to each sector of the pie chart **and** for drawing the pie chart.
  2. There is a tolerance of 2 degrees allowed for each sector.
  3. Use a protractor placed on the start line to measure each of your calculated angles around the circle. Each new line drawn becomes the start line for the next angle.
  4. You need to label each sector with the number of degrees and the data it represents.
- When drawing a histogram:
  1. Always use a pencil and a ruler.
  2. Label the axes.
  3. Make sure each bar is the same width.
  4. Remember: In a histogram, there is **no space** between the bars. They are joined together. This is because the data being represented is continuous data.
- In a stem and leaf diagram, the numbers **must** be in increasing order and you **must** have a key to achieve full marks.

INTRODUCTION

Exam Dictionary

Exam Terminology	What It Means
Calculate	Work out the value of something.
Comment/Give a reason for your answer	Use words (or mathematical symbols) to explain an answer. Must be clear and contain accurate reasoning.
Construct	Draw according to specific requirements. Accurate measurements are required. Construction lines must be shown clearly for full marks.
Describe	Give a detailed account in words.
Draw	Produce an accurate drawing.
Estimate the value of …/Find an approximate answer to …	Do not work out the exact answer. Use your own judgement regarding the level of rounding that is appropriate.
Evaluate	Find the value of.
Expand	Remove brackets and multiply out.
Explain your answer	Write a sentence to show how you got your answer or reached your conclusion.
Expand and simplify	Multiply out using distributive law and then group the like terms.
Express	Rewrite in another form.
Express, in terms of …	Use given information to write an expression using only the letter(s) given.
Factorise	Take out the common factor, or factorise into two brackets if a quadratic.
Fill in the table	Write all answers into the appropriate spaces in the table provided.
Graph	Present the data in the space provided. Draw clearly with a pencil. Use all available space so data can be clearly seen by the examiner. Label axes and make sure scales are equal and consistent. Position points accurately and label each point clearly.
Given	Use the specific data provided.
Give your answer in terms of π	must feature in your answer, e.g. 3π.
Give your answer in the form …	For full marks you must give your answer in the required format, e.g. specified number of decimal places or significant figures.
Hence	Use the previous answer to proceed.

Exam Terminology	What It Means
Hence, or otherwise	Use the previous answer but if you cannot see how to, you may use another method.
Identify	Provide an answer from a number of alternatives.
Interpret	Explain the meaning of something.
Justify	Write a sentence to show how you got your answer or reached your conclusion.
Make (x) the subject	Rearrange a formula.
Measure	Use a ruler or a protractor to measure a length or an angle.
Multiply out	Multiply out using distributive law.
Multiply out and simplify	Multiply out using distributive law and then group like terms.
Not to scale	This is often written in an exam question to discourage you from taking any measurements from the given diagram.
Plot	Accurately draw a graph.
Prove	A rigid algebraic or geometric proof is required.
Show that	Use words/numbers or algebra to show an answer.
Simplify	Group like terms or cancel down a fraction. Give your answer in the most compact, sufficient manner.
Solve	Get the answer using algebraic, numerical or graphical methods.
Use the graph	It's always worth indicating on the graph where the answer came from. No other method of obtaining the correct solution will be awarded full marks.
Using this, or otherwise	Use the previous answer but if you cannot see how to, you may use another method.
Verify	Prove that your answer is correct. In algebra, this may involve substituting your answer into the given equation and showing the result is correct.
Work out	Show all calculations in the space provided.
Write as a single fraction	You must find a common denominator and add/subtract the fraction.
Write down	You do not have to show any calculations, the correct answer is all that is required here.

	Mathematical Symbols
$\mathbb{N}$	Natural numbers – all positive whole numbers
$\mathbb{Z}$	Integers – all positive and negative whole numbers, including zero
$\mathbb{R}$	Real numbers – rational and irrational numbers
$\mathbb{Q}$	Rational numbers – any number that can be written as a fraction
$\mathbb{Q}/\mathbb{R}$	Irrational numbers – numbers that cannot be written as a fraction, e.g. $\sqrt{2}$ or π
$=$	Is equal to
$\neq$	Is not equal to
$\in$	Is an element
$\notin$	Is not an element of
$\subset$	Is a subset of
$\not\subset$	Is not a subset of
$\#$	The number of elements in a set
$\emptyset / \{\}$	The null set
$\cup$	Union of two sets
$\cap$	Intersection
$\setminus$	Set difference
$'$	Complement
$|AB|$	The length of the line AB; the distance between the points A and B
$>$	Greater than
$<$	Less than
$\geq$	Greater than or equal to
$\leq$	Less than or equal to
$\overrightarrow{AB}$	Direction of the translation
$\parallel$	Is parallel to
$\perp$	Is perpendicular to
$\angle A$	The angle A
△	Equilateral triangle
△	Isosceles triangle
$f(x)$	y
$\equiv$	Is congruent
$|||$	Is similar
∟	Right angle

SKILLS FOR EXAM SUCCESS MATHS

Chapter 1: Numbers

The topics and skills I will work on are:

Natural numbers:
- prime numbers
- composite numbers
- factors and highest common factors (HCFs)
- multiples and lowest common multiples (LCMs)
- prime factor decomposition

Integers

Rational numbers (numbers that can be written as fractions)
- Addition and subtraction of fractions
- Multiplication and division of fractions
- Ratios
- Decimals and rounding

Irrational numbers (numbers that cannot be written as fractions)

Real numbers
- Surds

Indices and index notation

Scientific notation

Section A: Natural Numbers (ℕ)

- Natural numbers (ℕ) are positive whole numbers {1, 2, 3, 4, 5, 6…}. The set is infinite.

- Natural numbers can be divided into two groups – prime numbers and composite numbers:
  - A prime number is a natural number that has only two factors – itself and 1.
    For example {2, 3, 5, 7, 11, 13, 17, 19…}
    Note: 1 is **not** a prime number.
  - Composite numbers are all other natural numbers. They are numbers that have **more than** two factors. For example {4, 6, 8, 9, 10, 12, 14, 15…}

- Factors are numbers that divide evenly into a certain number. For example, the factors of 6 are 1, 2, 3, 6.

- The highest common factor (HCF) of two numbers is the highest number that can divide evenly into both of these numbers.

For example: Find the highest common factor of 60 and 42.

60: 1, 2, 3, 4, 5, 6, 10, 12, 15, 20, 30, 60

42: 1, 2, 3, 6, 7, 14, 21, 42

HCF: 6

- A multiple of a natural number is the product of that number and an integer. For example, the multiples of 6 are 6, 12, 18, 24, 30, 36… The multiples of 5 are 5, 10, 15, 20, 25, 30…
- The lowest common multiple (LCM) of two numbers is the smallest multiple that both numbers share.

For example: Find the lowest common multiple of 4 and 5.

4: 4, 8, 12, 16, 20, 24, 28, …

5: 5, 10, 15, 20, 25, 30, …

LCM: 20

1. Look at the numbers 3, 4, 5, 9, 10, 12, 15, 18, 30.

   a) Which numbers are prime? _____

   b) Which numbers are composite? _____

   c) Which numbers have 6 as a factor? _____

   d) From the list, identify the factors of 20. _____

   e) From the list, identify the multiples of 5. _____

Exam Hint
Read the question carefully to make sure you answer it correctly.

Worked example
Write 126 as a product of its prime factors.

```
2 | 126
2 |  62
31|  31
         1
```

The smallest prime number that divides into 126 is 2.
The smallest prime number that divides into 62 is 2.
The smallest prime number that divides into 31 is 31.
No more divisions are possible.

126 as a product of its prime factors = 2 × 2 × 31 or $2^2 × 31$

This technique is called prime factor decomposition.

2. Shane got a new phone. He told his mother that his security code included the digits found from the prime factor decomposition of 2730. Find these digits and write them in ascending order.

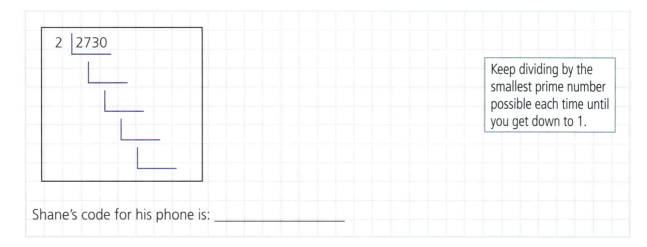

Keep dividing by the smallest prime number possible each time until you get down to 1.

Shane's code for his phone is: _____

3. Sophia is having a barbeque. She goes to her local supermarket to buy burgers and burger buns. Burgers are sold in packets of 8 and burger buns are sold in packs of 'Buy 10, get 2 free'.

   Sophia does not want excess burgers or buns. What is the smallest number of packets of burgers and buns she can buy so that each burger has a bun, with no waste?

Section B: Integers (ℤ)

- Integers (ℤ) are positive and negative whole numbers and include zero.
- The set of integers is infinite.
- Natural numbers are a subset of the integers.

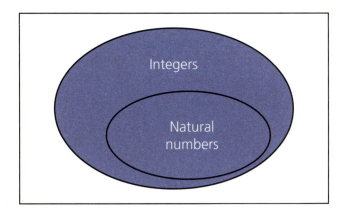

- When adding numbers, we move right along the number line.
- When subtracting numbers, we move left along the number line.

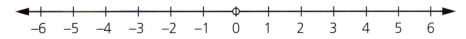

4. Work out the following:

   a) $4 + 5 =$

   b) $-3 + 7 =$

   c) $6 - 9 =$

   d) $-5 - 8 =$

Remember!

When we are multiplying or dividing integers, certain rules apply. It is extremely important that you know these rules as they play a vital role in every aspect of maths.

Rules of multiplication

$(+)(+) = +$

$(-)(-) = +$

$(+)(-) = -$

$(-)(+) = -$

Rules of division

$(+)/(+) = +$

$(-)/(-) = +$

$(+)/(-) = -$

$(-)/(+) = -$

CHAPTER 1: NUMBERS 3

5. Work out the following:

   a) $12 \div 4 =$

   b) $-16 \div 2 =$

   c) $-2 \times (-6) =$

   d) $4 \times (-5) =$

   e) $(-2)^2 =$

   f) $(-3)^3 =$

Section C: Rational Numbers ($\mathbb{Q}$)

$\mathbb{Q} = \{\frac{p}{q} \mid p \in \mathbb{Z}, q \in \mathbb{Z}, q \neq 0\}$

- A rational number is a number that can be expressed as a fraction of two integers, a numerator (p) and a denominator (q): $\frac{p}{q}$.

- Throughout maths you will be asked to give your answer in its simplest form. When dealing with fractions, this involves dividing the numerator and the denominator by the same number in order to write the fraction in its simplest form.

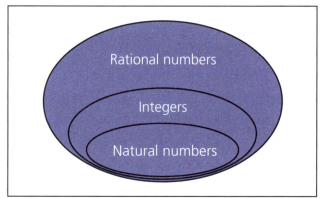

6. Write each of the following in its simplest form.

   a) $\frac{5}{10} =$

   b) $\frac{24}{30} =$

   c) $\frac{14}{49} =$

Addition and Subtraction of Fractions

- When adding and subtracting fractions, you need to first find a **common denominator** – this means having the same number on the bottom of all fractions.

- The common denominator is the LCM (we looked at this earlier in the chapter) of the denominators.

- The skills required to add, subtract, multiply and divide fractions are also necessary when working through other areas of maths, such as algebra, probability, applied measure and trigonometry, to name a few.

SKILLS FOR EXAM SUCCESS MATHS

Worked example

Write the following as a single fraction: $\frac{2}{3} + \frac{3}{5}$

Step 1: $\frac{}{15}$ — Find the common denominator.
Here the LCM of 5 and 3 is 15.
Once we convert both fractions to fifteenths, we can add them together.

Step 2: $\frac{5(2) + 3(3)}{15}$ — We multiplied the denominator 3 by 5 to convert to fifteenths. Now we must do the same to the numerator. We must multiply it also by 5, so we have 5(2).
+
We multiplied the denominator 5 by 3 to convert to fifteenths. Now we must do the same to the numerator. We must multiply it also by 3, so we have 3(3).

Step 3: $\frac{10 + 9}{15} = \frac{19}{15}$ — So the answer for this question would be $\frac{19}{15}$. This is called a top-heavy fraction.
If the answer was required as a mixed fraction then:
15 divides into 19 one time remainder 4.
So as a mixed fraction the answer would be $1\frac{4}{15}$.

7. Write the following as a single fraction in its simplest form:

$$3\frac{3}{5} - 2\frac{3}{4}$$

Firstly, convert the mixed fractions to top-heavy fractions:

$3\frac{3}{5} = \frac{18}{5}$ $(3 \times 5 + 3 = 15 + 3 = 18)$

$2\frac{3}{4} = \frac{11}{4}$ $(2 \times 4 + 3 = 8 + 3 = 11)$

You now follow the same steps as with the addition of fractions.

$$\frac{18}{5} - \frac{11}{4}$$

Step 1: $\frac{}{20}$

Step 2:

Step 3:

Exam Hint

I recommend that you write out each step clearly, so that if you do make a mistake, the examiner will be able to pinpoint exactly where your error is and deduct the minimum amount of marks.

CHAPTER 1: NUMBERS

Multiplication and Division of Fractions

Worked example

Work out $4\frac{1}{5} \times 2\frac{1}{7}$

Step 1: Write each fraction as a top-heavy fraction.

$\frac{21}{5} \times \frac{15}{7}$

No common denominator is required for multiplication or division.

Step 2: Multiply the numerators and multiply the denominators.

$\frac{21}{5} \times \frac{15}{7} = \frac{315}{35}$

Step 3: Simplify the answer if necessary.

$\frac{315}{35} = \frac{9}{1} = 9$

8. a) Work out $3\frac{2}{3} \times 4\frac{1}{5}$

Step 1: Write as top-heavy fractions.

Step 2: Multiply the numerators and multiply the denominators.

Step 3: Simplify.

b) Work out $2\frac{1}{4} \div 3\frac{3}{5}$

Step 1: Write as top-heavy fractions.

Step 2: Leave the first fraction as it is, change the division sign to multiplication and invert the second fraction. (KEEP. CHANGE. FLIP.)

Step 3: Multiply the numerators and multiply the denominators.

Step 4: Simplify.

Ratios

- A ratio compares sizes of quantities that make up a total.
- To convert a ratio into a fraction, add the two parts of the ratio together to make a total. This number will be the denominator.
- For example, a ratio of 2 : 3 becomes $\frac{2}{5}$ and $\frac{3}{5}$.

9. Sean makes a filling for his cakes using only butter and sugar. The ratio of the weight of butter to sugar is 4 : 8.

   One day Sean made a total of 3·6 kg of filling. Work out how many grams of sugar Sean used to make his filling.

Step 1: 4 + 8 = 12

Therefore $\frac{4}{12}$ of the mixture is butter and $\frac{8}{12}$ of the mixture is sugar.

Step 2: Convert 3·6 kg to grams.

Step 3: Find $\frac{8}{12}$ of the answer to step 2.

CHAPTER 1: NUMBERS

Decimals and rounding

- Throughout your exam you will be asked to give your answer to a specified number of decimal places or significant figures.

- If you are asked to give your answer to 2 decimal places, you need to have two numbers after the decimal point; if you are asked to give your answer to 3 decimal places, your final answer needs to have three numbers after the decimal point, and so on.

- If you are asked to give your answer to 2 decimal places, you look at the third digit. If it is 5 or more you round up; if it is less than 5, you round down. For example:

    2·347 = 2·35   2·342 = 2·34

- If you are asked to give your answer to 3 decimal places then you have to have three digits after the decimal point, for example: 3·1478 = 3·148.

- If you are asked to round to a number of significant figures, count the numbers from left to right, and then round off from there using the rules of rounding. When there are leading zeros (such as 0·003) don't count these as significant figures as they are there only to show how small the number is. For example, 3·1478 to 3 significant figures = 3·15.

Exam Hint

In the exam, you may be asked to round your answer. Marks will be deducted if the final answer is not rounded or is rounded incorrectly.

Take your time and read the questions carefully so that you are giving your answer in the required format.

Table of Decimal Places vs Significant Figures

	Significant figures	Decimal places
3·14159 Round to 3…	3·14	3·142
210·28901 Round to 2…	210	210·29
1567·326 Round to 2…	1600	1567·33

Worked example

- Write the following to 2 decimal places:

    2·367…………… We look at the *third digit* after the decimal point. Here it is 7. As this is 5 or greater, we round the second digit up.

    Answer: 2·37

- Write the following to 3 decimal places:

    4·5624 ………… We look at the *fourth digit* after the decimal point. Here it is 4. As this is less than 5, we leave the third digit as it is.

    Answer: 4·562

SKILLS FOR EXAM SUCCESS MATHS

10. Write the following to the number of decimal places specified.

   a) 5·784 =   (2 d.p.)

   b) 8·9767 =   (3 d.p.)

   c) 127·98789 =   (4 d.p.)

> **Worked example**
>
> a) Write 2356 to 2 significant figures.
>
> Two significant figures means that the final answer must include just two digits and then is rounded off from there.
>
> Step 1: We need to round this number to just two digits, with as many zeros as necessary.
>
> Step 2: 2356 is closer to 2400 than 2300.
>
> Answer = 2400
>
> b) Write 18 427 to 3 significant figures.
>
> Answer: 18 427 is closer to 18 400 than 18 500 so to 3 significant figures, 18 427 rounds to 18 400.

11. Write each of the following to 3 significant figures. (Remember, you can have as many zeros as necessary.)

   a) 0·002375 = 0·00238   *The fourth digit is 5, so we round up the third digit.*

   b) 0·0006543 =

   c) 19 678 =

   d) 22 344 =

CHAPTER 1: NUMBERS   9

Section D: Irrational Numbers ($\mathbb{Q}/\mathbb{R}$)

- Irrational numbers ($\mathbb{Q}/\mathbb{R}$) are numbers that cannot be expressed as fractions.
- Examples include π, $\sqrt{2}$, $\sqrt{3}$ and $\sqrt{5}$.

12. Which one of the following is not a rational number? Put a tick (✓) in the correct box.

    $3\frac{1}{7}$ ☐ $3\cdot142$ ☐ $\frac{22}{7}$ ☐ π ☐

    Give a reason for your answer.

Section E: Real Numbers ($\mathbb{R}$)

- Real numbers ($\mathbb{R}$) are **all** numbers – both rational and irrational.

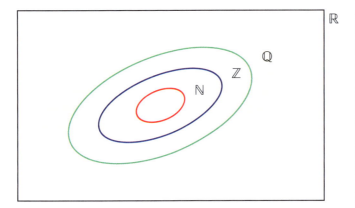

Remember!

Natural numbers ($\mathbb{N}$) are a subset of the integers ($\mathbb{Z}$).

The integers ($\mathbb{Z}$) are a subset of the rational numbers ($\mathbb{Q}$).

This means all natural numbers and integers are also rational numbers as they can be written over 1.

Rational numbers ($\mathbb{Q}$) and irrational numbers have nothing in common.

The real numbers ($\mathbb{R}$) are all numbers (natural numbers, integers, rational and irrational numbers).

Surds

- It is a common feature throughout maths, particularly in Algebra and Trigonometry, to be asked to give your answer in surd form.

- The laws of surds **are not** in the *Formulae and Tables* booklet and **need to be known by heart**. The three laws are:

    1. $\sqrt{ab} = \sqrt{a}\sqrt{b}$

    2. $\sqrt{a}\sqrt{a} = a$

    3. $\sqrt{\frac{a}{b}} = \frac{\sqrt{a}}{\sqrt{b}}$

Worked example

Express $\sqrt{98} + \sqrt{72} - \sqrt{32}$ in the form $a\sqrt{2}$.

Step 1: We factor each individual surd so that one of the factors will be $\sqrt{2}$.

$\sqrt{98} = \sqrt{49}\sqrt{2} = 7\sqrt{2}$

$\sqrt{72} = \sqrt{36}\sqrt{2} = 6\sqrt{2}$

$\sqrt{32} = \sqrt{16}\sqrt{2} = 4\sqrt{2}$

> There is a hint in the question as to how to find the solution. In the question we are told that the solution has to be in the form $a\sqrt{2}$. This tells us that we need to factor each of our surds so that one of their factors is $\sqrt{2}$.

Step 2: Rewrite the original equation.

$$\sqrt{98} + \sqrt{72} - \sqrt{32}$$
$$= 7\sqrt{2} + 6\sqrt{2} - 4\sqrt{2}$$
$$= 13\sqrt{2} - 4\sqrt{2}$$
$$= 9\sqrt{2}$$

13. Express $\sqrt{48} + \sqrt{243} - \sqrt{192}$ in the form $a\sqrt{3}$.

Exam Hint

Look for the clue in the question. In this case, you are asked to give your answer in the form $a\sqrt{3}$ therefore each surd must have $a\sqrt{3}$ as one of its factors.

Worked example

Without using a calculator, evaluate $(\sqrt{7} + 2\sqrt{3})(\sqrt{7} - 2\sqrt{3})$.

$(\sqrt{7} + 2\sqrt{3})(\sqrt{7} - 2\sqrt{3})$

$= \sqrt{7}(\sqrt{7} - 2\sqrt{3}) + 2\sqrt{3}(\sqrt{7} - 2\sqrt{3})$

$= 7 - 2\sqrt{3}\sqrt{7} + 2\sqrt{3}\sqrt{7} - 4\sqrt{3}\sqrt{3}$

$= 7 - 4(3)$

$7 - 12$

$= -5$

Step 1: Expand the brackets.
Write the contents of the second bracket out twice, split the contents of the first bracket.

Step 2: Multiply out
$\sqrt{7}\sqrt{7} = 7$
$\sqrt{7}(-2\sqrt{3}) = -2\sqrt{3}\sqrt{7}$
$2\sqrt{3}(\sqrt{7}) = 2\sqrt{3}\sqrt{7}$
$2\sqrt{3}(2\sqrt{3}) = 4\sqrt{3}\sqrt{3}$

Step 3: Simplify
$\sqrt{3}\sqrt{3} = 3$

14. Without using your calculator, evaluate $(3 - 3\sqrt{5})(3 + 3\sqrt{5})$.

Section F: Indices and Index Notation

- The following laws of indices can be found on page 21 of the *Formulae and Tables* booklet.

    Law 1: $a^p a^q = a^{p+q}$

    Law 2: $\dfrac{a^p}{a^q} = a^{p-q}$

    Law 3: $(a^p)^q = a^{pq}$

    Law 4: $a^0 = 1$

    Law 5: $a^{-p} = \dfrac{1}{a^p}$

    Law 6: $a^{\frac{1}{q}} = \sqrt[q]{a}$

    Law 7: $a^{\frac{p}{q}} = \sqrt[q]{a^p} = (\sqrt[q]{a})^p$

    Law 8: $(ab)^p = a^p b^p$

    Law 9: $\left(\dfrac{a}{b}\right)^p = \dfrac{a^p}{b^p}$

15. Evaluate the following using the laws of indices:

    a) $\dfrac{a^5 \times a^6}{a^4}$ (Laws 1 and 2)

    b) $32^{\frac{4}{5}}$ (Law 7)

16. Find the value of n so that the following is true for all $x \in R$.

    $\dfrac{x^4 \times x^n}{x^3} = x^6$

    Step 1: Law 1
    Step 2: Law 2
    Step 3: Use algebra to solve for n

SKILLS FOR EXAM SUCCESS MATHS

Section G: Scientific Notation

Worked example

Write 868 million in the form $a \times 10^n$ where $n \in \mathbb{Z}$ and $1 \leq a < 10$, $a \in \mathbb{R}$.

> n is an integer, which means it can be a positive or a negative whole number.
> a must be a number between 1 and 10.

868 million = 868 000 000

We need to convert 868 000 000 to a number between 1 and 10 to satisfy the above requirements.

Currently the decimal point is after the last zero of 868 000 000.

By moving the decimal point 8 places to the left, we convert the number to 8·68000000.

Write this in the required format: 8.68×10^8

To extend this question, write 8.68×10^8 correct to 2 significant figures. _____

17. Jupiter is approximately 778 600 000 km from the Sun. Write this number in the form $a \times 10^n$, where $n \in \mathbb{Z}$, a is $1 \leq a < 10$, correct to 2 significant figures.

CHAPTER 1: NUMBERS

Exam-Style Questions

Exam Hints

1. Attempt every part of every question. You cannot get marks for a blank space!
2. Highlight key words in the question. Make sure you understand what you are being asked.
3. Write the relevant formula from the *Formulae and Tables* booklet. The correct formula with some relevant substitution will get you marks.
4. If you are asked to give your answer in a particular format and are unsure of how to do this, do as much as you can.
5. Marks are not just given for the correct answer, they are also available for work completed along the way.

Question 1

a) Complete the table by writing Yes or No in each box to indicate whether each of the numbers is or is not an element of the natural numbers ($\mathbb{N}$), integers ($\mathbb{Z}$), rational numbers ($\mathbb{Q}$), irrational numbers ($\mathbb{Q} \setminus \mathbb{R}$) or real numbers ($\mathbb{R}$).

One example has been completed for you.

Number	Natural number	Integer	Rational number	Irrational number	Real number
$\sqrt{3}$	no	no	no	yes	yes
3					
−3					
$\frac{2\pi}{3}$					
0·33					

b) There are 120 students in first year.

i. $\frac{3}{5}$ of the year want to continue studying Science in second year.

How many students do not want to continue studying Science?

ii. 42 students opt to continue studying French in second year.

Calculate the percentage of the year who wish to continue studying French.

iii. The ratio of boys to girls in first year is 7 : 5.

How many girls are there in first year?

Question 2 – 2021

a) **i.** Write the numbers 3, 9 and 25 into the three empty boxes below to make the mathematical statement true. Use each number only once.

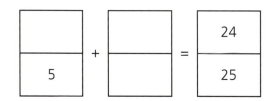

ii. Write the numbers 3, 5, 9 and 25 into the empty boxes below so that the difference between the two fractions is as large as possible. Use each number only once.

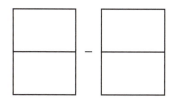

b) A positive whole number has exactly four factors. One of the factors is 9.

Work out the number.

Question 3

Use the properties of surds to show that $\sqrt{98} - \sqrt{18} + \sqrt{2}$ simplifies to $5\sqrt{2}$.

Exam Hint

Look for the clue in the question. You are asked to show your answer as $5\sqrt{2}$. This means that each of the surds must be factorised and one of the factors must be $\sqrt{2}$.

CHAPTER 1: NUMBERS **15**

Question 4

The diagram represents the sets of natural numbers (ℕ), integers (ℤ), rational numbers (ℚ) and real numbers (ℝ). Insert each of the following numbers into the correct place on the diagram:

−4 π $\frac{5}{6}$ $\sqrt{11}$ 0·45 4^{-2}

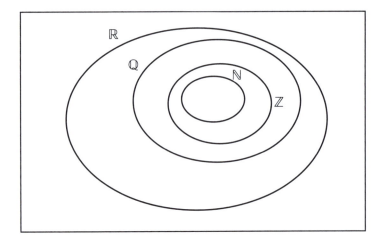

Remember!

Natural numbers are positive whole numbers.

Integers are positive and negative whole numbers.

Rational numbers are numbers that can be written as fractions.

Irrational numbers are numbers that cannot be written as fractions.

Summary review:

- You should now understand what natural numbers (ℕ), integers (ℤ), rational numbers (ℚ), irrational numbers (ℚ / ℝ) and real numbers (ℝ) are:

    - Natural numbers are positive whole numbers.

    - Integers include positive and negative whole numbers and zero.

    - Rational numbers are numbers that can be written as fractions.

    - Irrational numbers are numbers that cannot be written as fractions.

- You should be able to:

    - add, subtract, multiply and divide fractions

    - round your answers correctly – remember to give your answers to the specified number of decimal places or significant figures

    - write answers in the form $a \times 10^n$.

- You should also be familiar with:

    - the laws of indices – these are in the *Formulae and Tables* booklet (p. 21) so you do not need to remember them

    - the laws of surds – these are not in the *Formulae and Tables* booklet so you do need to remember them:
       1. $\sqrt{ab} = \sqrt{a}\sqrt{b}$ 2. $\sqrt{a}\sqrt{a} = a$ 3. $\sqrt{\frac{a}{b}} = \frac{\sqrt{a}}{\sqrt{b}}$

Scan this code to access a video with worked solutions for every question.

Chapter 2: Applied Arithmetic

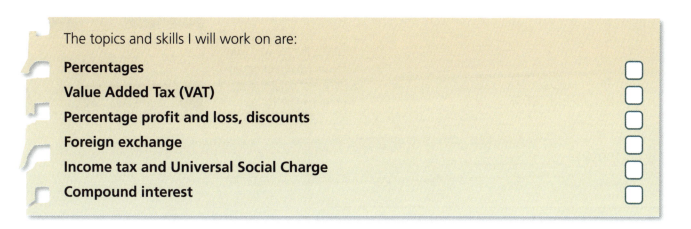

The topics and skills I will work on are:

Percentages

Value Added Tax (VAT)

Percentage profit and loss, discounts

Foreign exchange

Income tax and Universal Social Charge

Compound interest

Section A: Percentages

1. In 2020, the subscription to an online streaming service was €99·99. In 2022, the fee increased by 15%.

   Work out the cost of the service in 2022 and give your answer correct to 2 decimal places.

   Exam Hint
   Read each question carefully to make sure you are clear about what you are being asked to do.

   Step 1: Find 15% of €99·99.

   Step 2: Increase €99·99 by this amount.

   Step 3: Give your answer in the required format for the last few marks.

2. A travel agent receives a 5% commission for booking a family holiday to Disneyland. If she receives €150 commission, what is the cost of the holiday?

   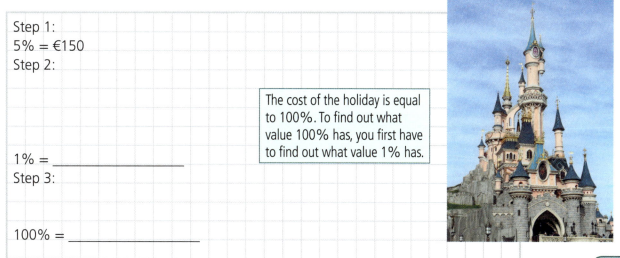

   Step 1:
   5% = €150
   Step 2:

   1% = _____
   Step 3:

   100% = _____

   The cost of the holiday is equal to 100%. To find out what value 100% has, you first have to find out what value 1% has.

CHAPTER 2: APPLIED ARITHMETIC **17**

3. John bought a new car in 2019. He sold it in 2022 for €25 000, at a loss of 40%. How much did the car cost him in 2019? Give your answer to the nearest cent.

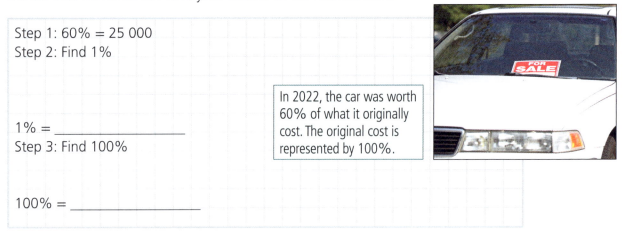

Step 1: 60% = 25 000
Step 2: Find 1%

1% = _____
Step 3: Find 100%

100% = _____

In 2022, the car was worth 60% of what it originally cost. The original cost is represented by 100%.

Section B: Value Added Tax (VAT)

- VAT is a tax that is added onto goods and services.

4. A computer game has a selling price of €79·99. This includes VAT at 21%.

   What is the price of the computer game before VAT is added? Give your answer to the nearest cent.

   Cost price + VAT = Selling price
   100% + 21% = 79·99
   121% = 79·99

   1% = _____

   100% = _____

   The selling price of an item is made up of its cost price and the VAT added together. We say that the cost price has a value of 100% and the VAT here has a value of 21%. This means the selling price has a value of 121%. When trying to find the cost price, we are trying to find 100% of the given amount.

5. Before VAT was added, an electricity bill was €184·65. After VAT was added, the bill was €209·58.

   Calculate the VAT rate.

   Give your answer correct to 2 significant figures.

   ### Exam Hint
   Marks will be deducted for failing to round your answer or for rounding incorrectly.

   Step 1: Calculate the amount of VAT that was added to the bill.

   Step 2:

   VAT rate = $\frac{\text{amount of VAT}}{\text{original amount}} \times 100\%$

   You need to remember this information as you will not be given it in the exam.

   Step 3: Answer in the required format.

18 SKILLS FOR EXAM SUCCESS MATHS

Section C: Percentage Profit

- Mark up is when the retailer adds a certain amount to the cost of goods to cover their expenses and to include a profit. It is calculated using $\frac{\text{Profit}}{\text{Cost price}} \times 100$.

- Margin represents the amount of a retailer's sales revenue that they get to keep as a profit, after all the costs have been subtracted. It is calculated using $\frac{\text{Profit}}{\text{Selling price}} \times 100$.

- You need to remember this information as you will not be given it in the exam.

6. A company bought a shipment of frames and art and craft materials to make personalised family picture frames for €150.

   They sold half of them at a 25% mark up and the other half at a 40% margin.

   Calculate the total selling price and therefore the total profit made by the company.

   Step 1: Half of the stock is sold at 25% mark up.
   €150 ÷ 2 = €75

   Step 2: 25% of €75 = _____
   Total selling price for this half of the stock: _____

   Step 3: The other half of the stock is sold at a 40% margin.
   60% = €75 (100% − 40% = 60%)
   1% = _____

   100% (Selling price for this half of the stock) = _____

   Step 4: Total selling price: _____

   Step 5: Total profit for the company: _____ (Total profit = Total selling price − Cost price)

CHAPTER 2: APPLIED ARITHMETIC

Section D: Foreign Exchange

7. Damien is a rugby player who is looking to buy a new sports vest. An American website is selling one for $299. A website in the UK is selling the same one for £250. On a day when €1 = US$1·08 and €1 = £0·89, from which website would you recommend Damien buy his sports vest?

To compare prices accurately, we need to convert them all to the same currency, in this case euros.

Step 1: We first need to calculate the value of 1 US dollar in euros.

$1·08 = €1

$1 = €$\frac{1}{1·08}$

Step 2: We then need to calculate the value of 299 dollars in euros.

$299 = €$\frac{1}{1·08}$ × 299

$299 = €276·8518

= €276·85   As we are talking about money, we round to 2 decimal places.

Step 3: We then need to repeat these steps to find the value of £250 in euros.

£0·89 = €1
£1 =

£250 =

Step 4: You are asked in the question 'which website would you recommend?' so don't forget to write a sentence by way of a conclusion to the question.
Conclusion: I would recommend that Damien buys the vest from
_____

8. Kacper is returning to Poland for a holiday. He buys 2345 złoty and the bank charges €512·50, which includes a service charge of 2·5%. What was the value of the euro in złoty, i.e. the exchange rate, on the day?

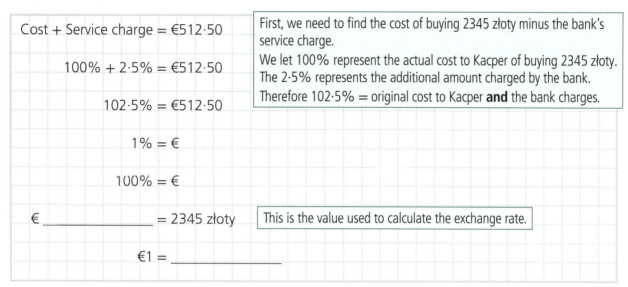

Section E: Income Tax

- Gross tax = Standard rate on all income up to the standard rate cut-off point (SRCOP) + Higher rate on all income above the SRCOP.

- Tax payable = Gross tax − Tax credits

- Net income = Gross income − Tax payable

Remember!
You need to remember this information as you will not be given it in the exam.

Worked example
Rose has a gross yearly income of €55 000. She has a standard rate cut-off point of €28 000. The standard rate of tax is 20% and the higher rate of tax is 42%. Her tax credits were €3560.

Exam Hint
In questions like these, it can be useful to highlight the key pieces of information.

Calculate her net income for the year.

Step 1: Calculate the gross tax payable (= Standard rate on all income up to the SRCOP + Higher rate on all income above the SRCOP).

To calculate the amount of money taxed at the higher rate, we have to subtract the SRCOP from the gross income (55 000 − 28 000).
= 20% of 28 000 + 42% of (55 000 − 28 000)
= 5600 + 42% of 27 000
= 5600 + 11 340
= 16 940

Step 2: Calculate the tax payable (= Gross tax − Tax credits).

16 940 − 3560

= 13 380

Step 3: Calculate the net income (= Gross income − Tax payable).

= 55 000 − 13 380

= €41 620

CHAPTER 2: APPLIED ARITHMETIC

Universal Social Charge (USC)

- This is another tax that people have to pay to the government.
- How does USC work? You have to pay:
    - 0·5% of any income you earn up to €12 012
    - 2% of any money earned between €12 013 and €21 295 (€9282)
    - 4·5% of any money earned between €21 296 and €70 044
    - 8% of any money earned over €70 045.

Remember!
You will be given the USC bands in the exam. You do not need to remember them.

Worked example

Calculate the USC payable on an income of €68 000.

Step 1: Calculate the amount of USC payable on the first €12 012 of the income.

   0·5% of 12 012 = €60·06

Step 2: Calculate the amount of USC payable on income between €12 013 and €21 295.

   2% of (21 295 − 12 013)

   2% of 9282 = €185·64

Step 3: Calculate the amount of USC payable on the remainder of the income.

   4·5% of (68 000 − 21 296)

   4·5% of 46 704 = €2101·68

Step 4: Total USC payable.

   €60·06 + €185·64 + €2101·68

   = €2347·38

Remember!
Any money earned between €21 295 and €70 044 is taxed at 4·5%.
This person earns €68 000.
To calculate the amount of money taxable at 4·5%:
€68 000 − €21 296 = €46 704 subject to tax at 4·5%

SKILLS FOR EXAM SUCCESS MATHS

9. Mark has a gross income of €60 000. His standard rate cut-off point is €36 000. The standard rate of tax is 22% and the higher rate of tax is 42%. His tax credit is €4200. He pays a pension contribution of €150 a month and health insurance of €55 per week.

Calculate Mark's:

a) gross tax

b) tax payable

c) USC

d) pension contribution per year

e) health insurance per year (assume a 52-week year)

f) net income.

Exam Hint

There is often a lot of information given in these questions. It can be useful to highlight the information that is relevant to you.

Step 1: Gross tax = Standard rate on all income up to the SRCOP + Higher rate on all income above the SRCOP

Step 2: Tax payable = Gross tax − Tax credits

Step 3: USC

Step 4: Pension contribution

Step 5: Health insurance

Step 6: Net income

Exam Hint

I recommend that you write out each step clearly so that if you do make a mistake, the examiner will be able to pinpoint exactly where your error is and deduct the minimum amount of marks.

Remember!

USC Bands

€0–€12 012 @ 0·5%

€12 013–€21 295 @ 2%

€21 296–€70 044 @ 4·5%

€70 045 + @ 8%

CHAPTER 2: APPLIED ARITHMETIC

Worked example

Maria paid €10 580 in tax for the year 2021. She had a tax credit of €2800 and a standard rate cut-off point of €31 000. The standard rate of tax was 22% and the higher rate of tax was 42%.

a) How much money was taxed at the higher rate of tax?

b) What was her gross income for the year?

Give all answers to the nearest euro.

a) Step 1: Gross tax = Standard rate on all income up to the SRCOP + Higher rate on all income above the SRCOP

   | Let x = amount of money taxed at 42%.

   22% of 31 000 + 42% of x

   6820 + 42% of x

   Step 2: Gross tax − Tax credits = Tax payable

   (6820 + 42% of x) − 2800 = 10 580 | 6820 − 2800 = 4020

   4020 + 42% of x = 10 580

   42% of x = 10 580 − 4020

   42% of x = 6560

   1% of x = 6560 ÷ 42

   1% of x = 156·19047

   100% of x = 156·19047 × 100

   x = €15 619·047

   x = €15 619 | Rounded to the nearest euro

b) Step 3: Gross income is income earned up to SCROP and income earned above SRCOP.

   31 000 + 15 619 = €46 619

10. Last year, Cian had a gross income of €58 000. His standard rate cut-off point was €33 400, the standard rate of tax was 22% and the higher rate of tax was 42%. His tax credits were €3000.

    This year, his gross income has increased. His tax rates, standard rate cut-off point and tax credits all remained the same.

    His tax payable is now €16 500.

    Calculate his new gross income correct to the nearest euro.

24 SKILLS FOR EXAM SUCCESS MATHS

Step 1: Gross tax = Standard rate on all income up to the SRCOP + Higher rate on all income above the SRCOP

Remember!
Let x = amount of money taxed at 42%.

Step 2: Gross tax − Tax credits = Tax payable

Step 3: Gross income = income earned up to the SCROP + income earned above the SRCOP

11. Kathy has a gross income of €60 000. Her total tax payable was €12 500.

    The SRCOP is €29 000. The standard rate of tax is 22% and the higher rate of tax is 42%. Calculate her tax credits for the year.

    Step 1: Gross tax = Standard rate on all income up to the SRCOP + Higher rate on all income above the SRCOP

    Step 2: Gross tax − Tax credits = Tax payable
    Rearrange this formula: Gross tax − Tax payable = Tax credits

Section F: Compound Interest

- Compound interest is interest that is calculated not only on the initial principal (P) but also on the accumulated interest of previous years. It is calculated using the formula $F = P(1 + i)^t$, where F = final value, P = principal, i = interest rate that is charged per year and t = time that the money was invested/borrowed for.

- Depreciation is the loss of an item's value over time. It is calculated using the reducing balance method, $F = P(1 − i)^t$, where F = later value and P = initial value.

- These formulae are on page 30 of the *Formulae and Tables* booklet, so you do not have to remember them.

CHAPTER 2: APPLIED ARITHMETIC   25

12. Kate borrowed €15 000 from her local credit union to buy a new car. She agreed to pay back the loan over 5 years, at an interest rate of 4·5%.

    Calculate the amount Kate had to repay over 5 years. Give your answer to the nearest euro.

$F = P(1 + i)^t$	F = Final amount to be repaid
	P = Amount of money borrowed (€15 000)
	i = Interest rate. This is usually converted from a percentage to a decimal and substituted into the formula as a decimal (4·5% = 0·045)
	t = time the loan was taken out for (5 years)
$F = 15\,000\,(1 + 0.045)^5$	
$F =$	

13. Maura borrowed €40 000 from her local bank. The money was borrowed for 3 years, with an interest rate of 6% per annum.

    She also agreed to pay off €5000 at the end of each year. How much was outstanding in the loan after the third repayment?

    Year 1:
    $F = P(1 + i)^t$
    $F = 40\,000\,(1 + 0.06)^1$
    $F = 42\,400$
    Repayment $- 5000$
    $F = 37\,400$

    Year 1:
    F = final amount $P = 40\,000$ $i = 6\% = 0.06$
    (The interest rate is substituted into the formula as a decimal and $t = 1$ as we are working one year at a time.)
    At the end of the first year, €5000 is repaid to the bank.

    Year 2:
    $F = P(1 + i)^t$
    $F =$
    $F =$
    Repayment $- 5000$
    $F =$

    Year 2:
    F = final amount $P = 37\,400$ $i = 6\% = 0.06$
    At the end of the end of the second year, €5000 is repaid to the bank.

    Year 3:
    $F = P(1 + i)^t$
    $F =$
    $F =$
    Repayment $- 5000$

    Year 3:
    F = final amount $P =$ $i =$
    At the end of the third year, €5000 is repaid to the bank.

    Total amount owed to the bank after 3 years: _____

14. Kieran has €10 000 and would like to invest it. A special savings account is offering a rate of 3% for the first year, 5% for the second year and 7% for the third year.

    Tax at a rate of 18% is deducted from the interest earned every year. How much will the investment be worth at the end of the 3 years?

    Step 1:
    $$F = P(1 + i)^t$$
    $$F = 10\,000(1 + 0.03)^1$$
    $$F = 10\,300$$

    Step 2:
    Interest earned in year 1 = F – Amount at the start of year 1:
    10 300 – 10 000 = 300

    Step 3:
    18% of 300 = 54

    Step 4:
    Net interest = Interest earned – Tax
      300 – 54 = 246
    Amount at the end of year 1 = Initial amount at the start of the year + Net interest
       10 000 + 246
         10 246

    > You have to calculate the interest earned before any tax is paid.
    > €300 is earned in interest. Tax is only paid on the interest earned.
    > We need to calculate 18% of 300 and subtract this value from 300 to find the amount of interest added to the investment.

    Year 2: $F = P(1 + i)^t$
    $F =$
    $F =$
    Interest earned in year 2 = F – Amount at the start of year 1:
    18% of _____ =

    Net interest = Interest earned – Tax

    > Year 2: $P = 10\,246$ $i = 5\%$

    Amount at the end of year 2 = Initial amount at the start of year 2 + Net interest

    Year 3:

15. €3000 was invested for a year at x% interest rate. A tax of 20% was deducted from the interest earned each year. At the end of the first year, the investment amounted to €3060 after the 20% tax was deducted.

    Calculate the interest rate for the year.

    > The amount of interest earned was €60 **after** 20% tax had been deducted.
    > 80% = €60

16. A sum of money was invested for 2 years. After 2 years, the money amounted to €3244·50. The interest rate for the first year was 5% and for the second year was 3%.

    Calculate the amount of money invested.

    Step 1: At the end of the second year, the investment amounted to €3244·50. This included the amount of money in the account at the start of the second year and 3% interest.

    > Amount in the account at the start of the second year = 100%
    > Interest rate = 3%
    > Total amount = 103%

    103% = 3244·50
    1% = 31·50
    100% = 3150

    The amount of money in the account at the start of the second year is equal to the amount in the account at the end of the first year.

    This amount equals the original investment and 5% interest.

    Step 2:
    105% = 3150
    1% =

    > Amount in the account at the start of the second year = 100%
    > Interest rate = 5%
    > Total amount = 105%

    100% =

    This is the value of the original investment.

SKILLS FOR EXAM SUCCESS MATHS

Exam-Style Questions

Exam Hints

1. Attempt every part of every question. You cannot get marks for a blank space!

2. Highlight key words in the question. Make sure you understand what you are being asked.

3. Write the relevant formula from the *Formulae and Tables* booklet. The correct formula with some relevant substitution will get you marks.

4. If you are asked to give your answer in a particular format and are unsure of how to do this, do as much as you can.

5. Marks are not just given for the correct answer, they are also available for work completed along the way.

Question 1 – 2022

a) Jane buys a laptop online for $699, plus a shipping cost of $30.

The exchange rate is $1 = €0·90.

Work out in euros the total cost to Jane of buying the laptop online.

Don't forget to include the cost of the shipping.

b) Jane has a gross annual income of €56 000.

Jane pays income tax on her gross income at a rate of 20% on the first €44 300, and 40% on the balance.

i. Work out Jane's annual income tax at each of these two rates (20% and 40%).

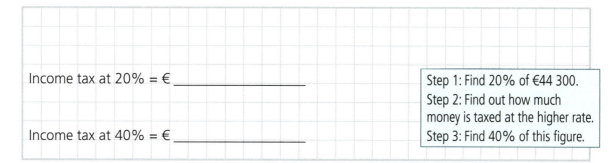

Income tax at 20% = € _____

Income tax at 40% = € _____

Step 1: Find 20% of €44 300.
Step 2: Find out how much money is taxed at the higher rate.
Step 3: Find 40% of this figure.

CHAPTER 2: APPLIED ARITHMETIC   29

(ii) Jane has annual tax credits of €3300.

Work out Jane's annual take-home pay.

Tax payable = Gross tax − Tax credits
Net income = Gross income − Tax payable

Question 2 – 2021

Millie bakes cakes and sells them at the local market.

Millie is buying flour at her local shop. The shop has two different special offers:

Special Offer A	Special Offer B
1 kg bags: €3·50 each	1·5 kg bags: €5 each
Special offer:	Special offer:
3 bags for the price of 2	20% off

Exam Hint

In the exam, there will be marks for each correct step that you make. So even if you can only do one part of the question, try that much! Some marks are better than no marks!

a) Millie wants to buy 6 kg of flour. Work out which offer, A or B, will give her the better value.

Better value

(tick (✓) **one** box only) Offer A   Offer B
                              ☐   ☐

Step 1: Work out the cost of buying 6 kg of flour with offer A.
Step 2: Work out the cost of 6 kg of flour with offer B.

b) Millie sells each cake for €7·50.

This gives her a profit of 20%.

Work out how much it costs Millie to make each cake.

€7·50 **includes** the cost to make the cake (100%) **and** the profit (20%). You need to calculate the cost (100%) to make the cake.

c) Millie has €3000 in a special savings account.

It has an interest rate of 2·5% per year for 4 years, compounded annually.

She does not put any money in or take any money out of the account over the 4 years.

Work out the total amount in the account after 4 years.

Give your answer correct to the nearest cent.

Remember: $F = P(1+i)^t$

Question 3

Linked Question: Numbers, Arithmetic, Distance/Time/Speed and Trigonometry

a) Sinéad is 14 years old and Veronika is 16 years old. They spent their summer babysitting. They earned €1200. They decided to divide the money in the ratio of their ages.

How much did each girl get?

Add $14 + 16 = 30$
So Sinéad gets $\frac{14}{30}$ of the money
and Veronika gets $\frac{16}{30}$ of the money.

b) Sinéad wants to go on holiday in 2 years' time. She decides to invest her wages in a savings account that offers an interest rate of 2% for the first year and 2·5% for the second year.

She has to pay 18% tax on any interest earned.

How much money will be in the account in 2 years' time?

Give your answer to the nearest euro.

Exam Hint

Marks will be deducted for failing to round your answer or rounding incorrectly.

Remember: $F = P(1+i)^t$

CHAPTER 2: APPLIED ARITHMETIC

c) Veronika is going to America. She wants to exchange her wages for dollars.

The exchange rate is €1 = $1·09

Calculate how much, in dollars, Veronika will receive.

d) A taxi to the airport costs €60. This is inclusive of VAT at 23%.

Calculate the fare before VAT was added. Give your answer to 2 decimal places.

> The €60 includes the original cost (100%) **and** the VAT at 23%. So €60 = 123%.

e) The distance from Shannon to New York is 4960 km.

The plane flies at an average speed of 805 km/hr.

Calculate the flight time from Shannon to New York.

Give your answer in hours and minutes, correct to the nearest minute.

> Remember: 15 minutes is $\frac{1}{4}$ of an hour or 0·25.

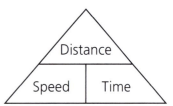

f) The aeroplane is cruising at an altitude of 3500 m when it begins its descent into JFK Airport. To land safely, the aeroplane makes an angle of 3° with the ground.

Using trigonometry, calculate the distance (at ground level) from JFK at which the pilot should begin his descent.

Give your answer to the nearest kilometre.

SOH $\sin \theta = \frac{\text{opposite}}{\text{hypotenuse}}$

CAH $\cos \theta = \frac{\text{adjacent}}{\text{hypotenuse}}$

TOA $\tan \theta = \frac{\text{opposite}}{\text{adjacent}}$

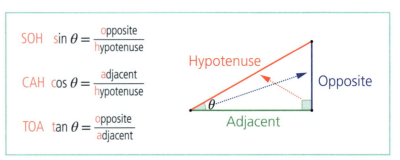

32 SKILLS FOR EXAM SUCCESS MATHS

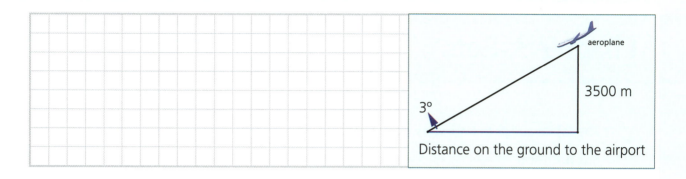

Summary review:

- You need to remember the following formulae as they are not in the *Formulae and Tables* booklet:

  - VAT rate = $\dfrac{\text{Amount of VAT}}{\text{Original amount}} \times 100\%$

  - Mark up = $\dfrac{\text{Profit}}{\text{Cost price}} \times 100\%$

  - Margin = $\dfrac{\text{Profit}}{\text{Selling price}} \times 100\%$

  - Gross tax = Standard rate on all income up to the SRCOP + higher rate on all income above SRCOP

  - Tax payable = Gross tax − Tax credits

  - Net income = Gross income − Tax payable

  - Selling price = Cost price + VAT

- The formula for compound interest and depreciation is on page 30 of the *Formulae and Tables* booklet.

- You do not need to remember tax rates, USC bands or VAT rates. These will be given to you in the exam.

Scan this code to access a video with worked solutions for every question.

CHAPTER 2: APPLIED ARITHMETIC   33

Chapter 3: Sets

The topics and skills I will work on are:

- Set notation
- Venn diagrams
- Union and intersection
- Complement of a set
- Set difference
- Numerical problems for two sets
- Numerical problems for three sets

Section A: Set Notation

Symbol	Meaning	Example
$\in$ *	Is an element of	$4 \in \{1, 2, 3, 4\}$
$\notin$	Is not an element of	$5 \notin \{1, 2, 3, 4\}$
$\subset$ *	Is a subset of	$\{1, 2\} \subset \{1, 2, 3, 4\}$
$\not\subset$	Is not a subset of	$\{4, 5\} \not\subset \{1, 2, 3, 4\}$
$=$	Is equal to	$\{1, 2, 3, 4\} = \{2, 1, 3, 4\}$
$\neq$	Is not equal to	$\{1, 2, 3, 4, 5\} \neq \{1, 2, 3, 4\}$
#	Cardinal number/ number of elements in	$\#\{1, 2, 3, 4\} = 4$
$\cup$ *	Union	$\{1, 2, 3, 4\} \cup \{3, 4, 5, 6\} =$ $\{1, 2, 3, 4, 5, 6\}$
$\cap$ *	Intersection	$\{1, 2, 3, 4\} \cap \{3, 4, 5, 6\} =$ $\{3, 4\}$
$\mathbb{U}$	Universal set	All of the elements
\ *	Set difference	$\{1, 2, 3, 4\} \setminus \{3, 4, 5, 6\} = \{1, 2\}$
'	Complement	$U = \{1, 2, 3, 4, 5, 6, 7, 8\}$ $A = \{1, 2, 3, 4\}$ $A' = U \setminus A = \{5, 6, 7, 8\}$
{ } or $\varnothing$	Null set	No elements in this set

* These are provided on page 23 of the *Formulae and Tables* booklet and do not need to be remembered.

34 SKILLS FOR EXAM SUCCESS MATHS

1. $\mathbb{U} = \{a, b, c, d, e, f, g, h, i, j, k\}$

   $A = \{a, b, c, d, e\}$

   $B = \{d, e, f, g, h, i\}$

   a) Represent this information on the Venn diagram below.

   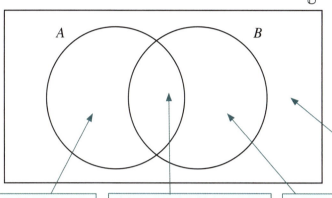

   Step 1: Fill in the elements that go into the intersection first.
   Step 2: Fill in the elements that are in set A only.
   Step 3: Fill in the elements that are in set B only.
   Step 4: Add any remaining elements into the region outside sets A and B but within the universal set.

   2. Elements that are in set A only.
   1. Intersection — the elements that are in set A and set B.
   3. Elements that are in set B only.
   4. Elements remaining in the universal set.

   b) Are the following statements true or false?

   i. $b \in A$ _____ ii. $f \notin B$ _____

   iii. $\{\} \subset A$ _____ iv. $\{d, e, f\} \not\subset B$ _____

   v. $\#A = 5$ _____ vi. $A = B$ _____

   c) List the elements of the following:

   i. $A \cup B =$ _____ Elements of A and B together

   ii. $A \cap B =$ _____ Elements in both A and B

   iii. $A' =$ _____ Universal set − Elements of set A

   iv. $B' =$ _____ Universal set − Elements of set B

   v. $A \backslash B =$ _____ Set A − Elements of set B

   vi. $B \backslash A =$ _____ Set B − Elements of set A

   vii. $(A \cup B)' =$ _____ Universal set − Elements of $A \cup B$

   viii. $(A \cap B)' =$ _____ Universal set − Elements of $A \cap B$

   ix. $A' \cup B =$ _____ (Universal set − Elements of set A) + set B

   x. $(A \cup B)' \cup (A \cap B)'$ _____

   1. Universal set − Elements of $A \cup B$
   2. Universal set − Elements of $A \cap B$
   3. Elements from step 1 + Elements from step 2

CHAPTER 3: SETS

Section B: Venn Diagrams

2. Shade the relevant regions in the Venn diagrams below.

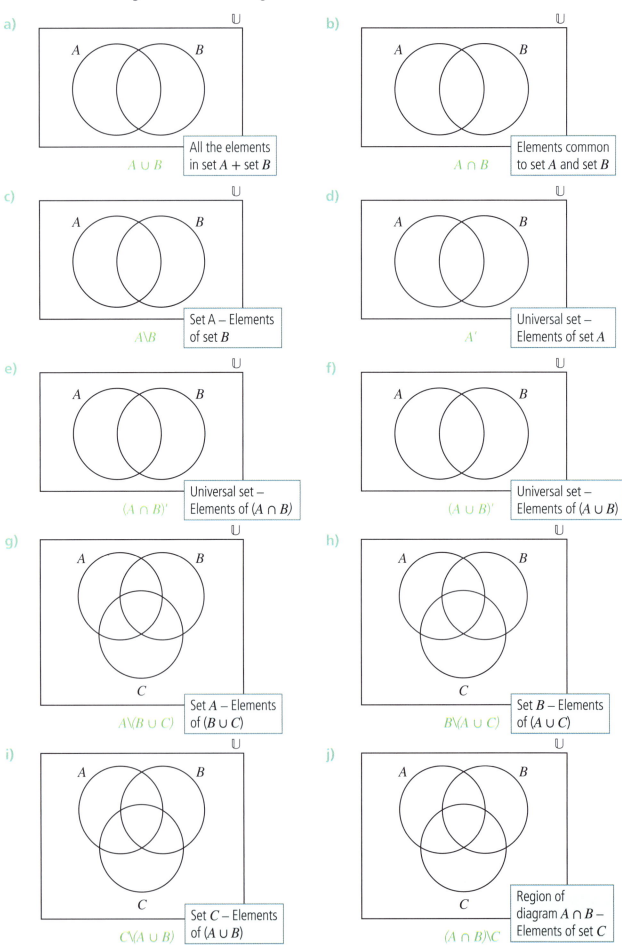

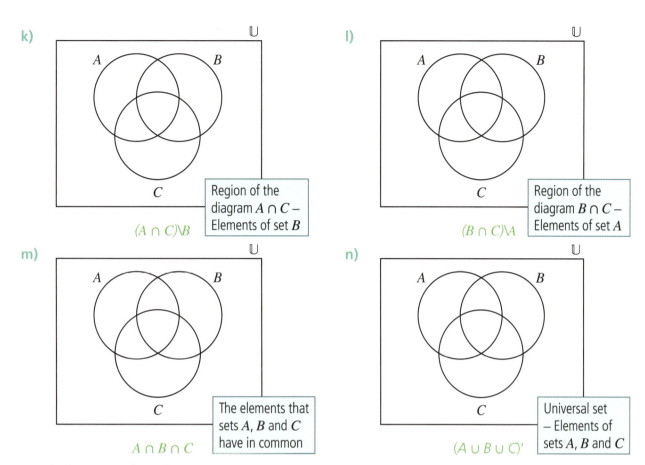

3. $\mathbb{U} = \{a, b, c, d, e, f, g, h\}$, $A = \{a, b, c, d\}$, $B = \{c, d, e, h\}$ and $C = \{a, c, e, g\}$.

   a) Represent these three sets on a Venn diagram.

   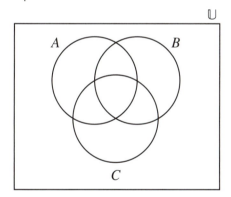

   Commutative – changing the order of the operations does not change the result.
   Associative – changing the grouping does not affect the result.

   b) Show that $(A \cup B) \cup C$ and $A \cup (B \cup C)$ are commutative.

   c) Show that $(A \cap B) \cap C$ and $A \cap (B \cap C)$ are associative.

   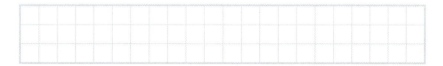

   d) Show that $(A \setminus B) \setminus C$ and $A \setminus (B \setminus C)$ are not commutative.

CHAPTER 3: SETS

Section C: Numerical Problems for Two Sets

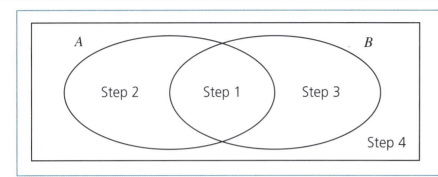

Step 1: Fill in $A \cap B$ (i.e. both).
Step 2: Fill in $A \backslash B$ (i.e. A only).
Step 3: Fill in $B \backslash A$ (i.e. B only).
Step 4: Fill in $\mathbb{U} \backslash (A \cup B)$ (i.e. neither).
Step 4: Check your data by adding all the terms within the Venn diagram and making sure it is the same as the total number given in the question.

4. In a class of 38 pupils, 23 study Physics, 19 study Chemistry and 8 study both.

Worked example

a) Represent this information on a Venn diagram.

Step 1: Fill in the intersection first. These are the pupils who study both Chemistry and Physics.

Step 2: Fill in the pupils who study Physics only.
Physics only: 23 pupils in total study Physics. 8 of these have been accounted for in the intersection.
Physics only: $23 - 8 = 15$

Step 3: Fill in the students who study Chemistry only.
Chemistry only: 19 students study Chemistry. 8 of these students have already been accounted for in the intersection.
Chemistry only: $19 - 8 = 11$

Step 4: Find the number of students who studied neither subject.
$38 - (15 + 8 + 11) = 4$

Physics only | Both Physics and Chemistry | Chemistry only | Neither Chemistry nor Physics

38 SKILLS FOR EXAM SUCCESS MATHS

b) Find the number of students who study:

    i.   Physics only   _____

    ii. Chemistry only   _____

    iii. Neither Physics nor Chemistry   _____

    iv. One subject only   _____

5. In a survey of 80 people, 54 had watched Netflix the previous night and 28 had watched Now TV.

   If 24 people had watched neither of these streaming channels, how many people had watched both?

   Represent this information on a Venn diagram.

Exam Hint

Even if you are not sure how to completely fill in the Venn diagram, fill in the regions that you do know. There will be marks for each step and some marks are better than no marks!

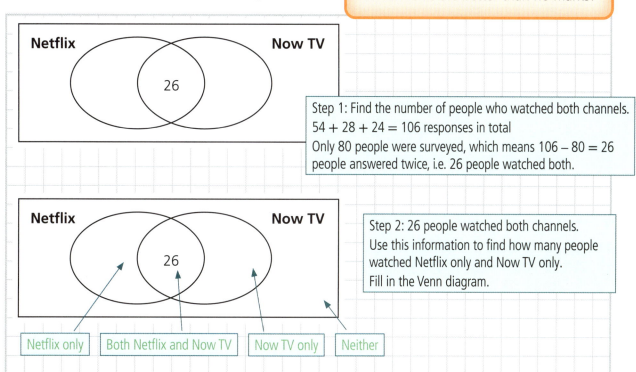

Step 1: Find the number of people who watched both channels.
54 + 28 + 24 = 106 responses in total
Only 80 people were surveyed, which means 106 − 80 = 26 people answered twice, i.e. 26 people watched both.

Step 2: 26 people watched both channels. Use this information to find how many people watched Netflix only and Now TV only. Fill in the Venn diagram.

6. $\#\mathbb{U} = 32$, $\#A = 16$, $\#B = 12$, and $\#A \cup B' = 10$.

   a) Use this information to calculate $\#(A \cap B)$.

   b) Represent the information on a Venn diagram.

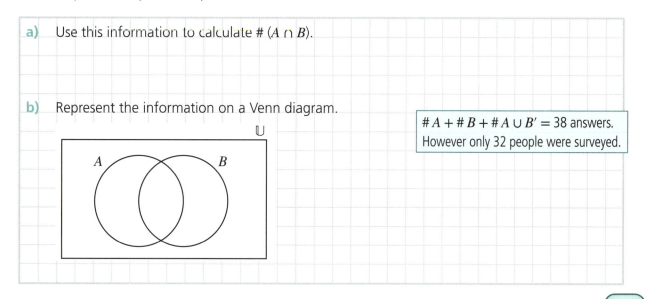

$\#A + \#B + \#A \cup B' = 38$ answers. However only 32 people were surveyed.

CHAPTER 3: SETS

c) Using your Venn diagram, work out:

   i. # (A\B) = _____

   ii. # (B\A) = _____

   iii. # (A ∩ B)' = _____

7. # U = 32, # (A\B) = twice # (A ∩ B), # B = 12 and # (A ∪ B)' = $\frac{1}{2}$ # B.

   Let *x* represent the number of elements in (A ∩ B). Fill the information into a Venn diagram and evaluate the value of *x*.

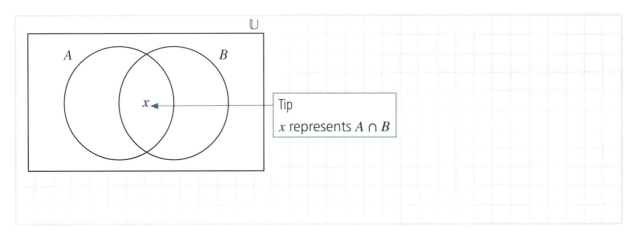

Section D: Numerical Problems for Three Sets

8. The Venn diagram below represents the subjects taken by 42 students in fifth year. The subjects are Accounting, Biology and Chemistry. Answer the questions that follow.

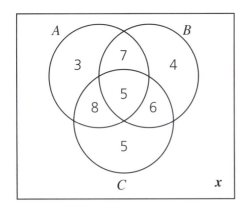

   How many students study:

   a) Accounting only _____

   b) All three subjects _____

   c) None of these three subjects _____

   d) Accounting and Biology but not Chemistry _____

   e) At least two subjects _____

   f) Only one subject _____

   g) Chemistry and Biology _____

   h) Only Biology _____

9. Each of the students in sixth year in a particular school has WhatsApp (W), Instagram (I) and/or Snapchat (S). The numbers who have each app are as follows:

36 students have WhatsApp

40 students have Instagram

54 students have Snapchat

14 students have WhatsApp and Instagram

24 students have Instagram and Snapchat

x students have WhatsApp and Snapchat, but not Instagram

8 students have all three apps.

Exam Hint

When filling in a Venn diagram, I advise always starting in the middle, here with $W \cap I \cap S$.

Worked example

a) Fill in the Venn Diagram.

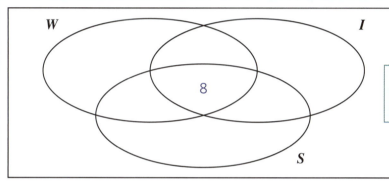

Step 1: Start by filling in the region $W \cap I \cap S$ – the students who have all three apps.

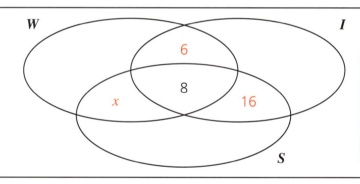

Step 2: Fill in the following three regions: those who have WhatsApp and Snapchat only, WhatsApp and Instagram only and Snapchat and Instagram only.
$(W \cap S) \backslash I = x$
$(W \cap I) \backslash S = 14 - 8 = 6$
$(S \cap I) \backslash W = 24 - 8 = 16$

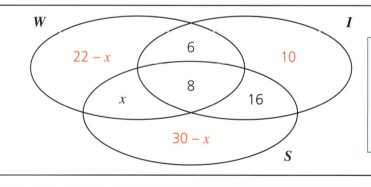

Step 3: Fill in the last three regions: those who have WhatsApp only, Instagram only and Snapchat only.
WhatsApp only: $36 - (6 + 8 + x) = 22 - x$
Instagram only: $40 - (6 + 8 + 16) = 10$
Snapchat only: $54 - (16 + 8 + x) = 30 - x$

CHAPTER 3: SETS

b) Given that there were 80 students in the year, solve for x.

$22 - x + 6 + 8 + x + 10 + 16 + 30 - x = 80$

c) Find the probability that a student chosen at random has:

i. all three apps =

ii. at least two apps =

iii. Snapchat only =

iv. one app only =

Give each answer as a fraction in its simplest form.

10. A survey was carried out and 31 people participated. They were asked if they had travelled to Tenerife, Disneyland Paris or London. The results were as follows:

18 had been to Tenerife
13 had been to Disneyland
17 had been to London
7 had been to Tenerife and Disneyland
4 had been to Disneyland and London
8 had been to Tenerife and London
x had been to all 3 destinations.

a) Represent this information in terms of x on a Venn diagram.

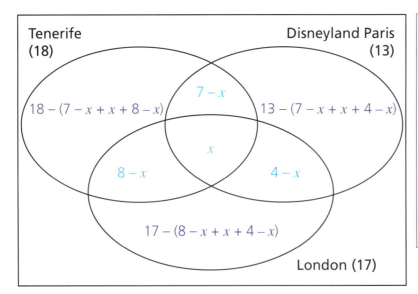

Step 1: I would advise starting in the middle with the people who had been to all three destinations.
Step 2: Fill in those who had been to Tenerife and Disneyland only, Tenerife and London only and London and Disneyland only.
Step 3: Finally, fill in those who had been to Tenerife only, Disneyland only and London only.

b) Solve for x.

SKILLS FOR EXAM SUCCESS MATHS

Exam-Style Questions

Exam Hints

1. Attempt every part of every question. You cannot get marks for a blank space!
2. Highlight key words in the question. Make sure you understand what you are being asked.
3. If you are asked to give your answer in a particular format and are unsure of how to do this, do as much as you can.
4. Marks are not just given for the correct answer, they are also available for work completed along the way.

Question 1 – 2022

Linked Question: Sets, Numbers, Arithmetic

80 students in a group were asked what they had done during their summer holidays.

Some of the students got a job (*J*), some went on holidays (*H*), some did both, and some did neither.

$\frac{1}{5}$ of the students in the group did neither.

25% of the students got a job.

Of those students who got a job, half also went on holidays.

Work out the total number of students in the group who went on holidays.

Use the Venn diagram below to help answer the question.

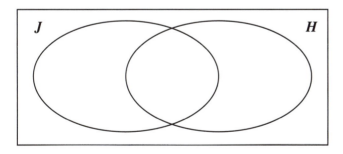

Step 1: Find $\frac{1}{5}$ of 80.
Step 2: Find 25% of 80 to find the total number of students who got a job.
Step 3: Find half of 25% of 80 to find the number of students who got a job and also went on holiday.
Step 4: Use the data to find how many people went on holidays only.

CHAPTER 3: SETS 43

Question 2 – 2021

a) *A*, *B* and *C* are sets.

Complete the tables below by shading in each of the given sets in the Venn diagram.

Exam Hint
When shading, always use a pencil so that if you make a mistake, you can easily rub it out.

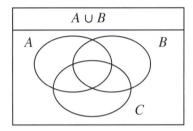
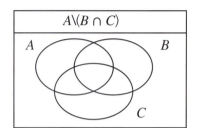

b) *P* and *Q* are two other sets. *P* is a subset of *Q*.

Remember!
Subset means that all the elements of *P* are also in *Q*.

i. Write an X in the region of the Venn diagram below which must contain no elements.

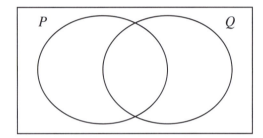

ii. Place a tick (✓) in the correct box to show which statement must be true. Tick one box only. Explain your answer.

$\#P \leq \#Q$ $\#P = \#Q$ $\#P \geq \#Q$

Exam Hint
Even if you are unsure how to phrase the explanation, tick the box you think is correct! There are marks available for each correct step along the way.

Question 3

Linked Question: Sets, Algebra, Numbers, Probability, Statistics, Arithmetic

55 people were asked if they ever had Covid-19, flu or chickenpox. The results were as follows:

31 had Covid-19
20 had flu
25 had chickenpox only
9 had Covid-19 and flu
10 had chickenpox and Covid-19
7 had chickenpox and flu
2 people had none of the three diseases.

Let *x* represent the number of people who had all three diseases.

Exam Hint
There are marks for each step in the question so even if you are unsure how to complete the Venn diagram fully, insert the data that you do know, even if that is just one or two terms.
You never know what Low Partial Credit will be allocated for. Some marks are better than no marks.

44 SKILLS FOR EXAM SUCCESS MATHS

a) Fill in the Venn diagram in terms of x.

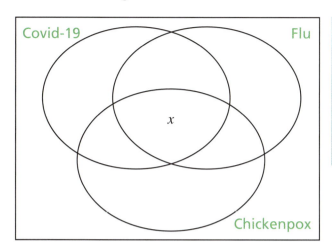

Step 1: Start in the middle with the people who had all three diseases.
Step 2: Fill in those who had Covid-19 and flu only, Covid-19 and chickenpox only and chickenpox and flu only.
Step 3. Fill in those who had Covid-19 only, flu only, chickenpox only and those who had none of the three diseases.

b) Use your data to solve for x.

Even if you feel the answer you got in part (a) is incorrect, use the data that you got to try to solve for x. You might get marks for using your incorrect data correctly.

c) Calculate the probability that a person chosen at random from the group had:

　i.　Flu

　ii.　At least two diseases

　iii.　Covid-19 only.

Give each answer as a fraction in its simplest form.

Exam Hint

You are asked to give each answer in its simplest form. You need to complete this last step to get all the marks available for this question.

i. Flu =

ii. At least two diseases =

iii. Covid-19 only =

CHAPTER 3: SETS 45

d) Calculate the ratio of people who had Covid-19 only to the people who had chickenpox only. Give your answer in its simplest form.

e) The following is data taken from the CSO website, relating to the uptake of the Covid-19 vaccine in 2021.

Vaccine uptake rate by age group by sex for employees aged 18 and over in 2021

Age group	Male uptake %	Female uptake %
18–24 years	78%	84%
25–44 years	78%	83%
45–64 years	89%	93%
65 years and over	94%	94%

Draw a suitable chart to illustrate this data.

Remember!
When drawing your chart:
- use a pencil
- use all the space available to you so everything is clear to the examiner
- label all data clearly.

f) If there were approximately 970 000 females aged between 25 and 44 living in Ireland in 2021, how many of these were vaccinated against Covid-19?

Give your answer in the form $a \times 10^n$, where $n \in \mathbb{Z}$, a is $1 \leq a < 10$, correct to 3 significant figures.

Exam Hint
Even if you can't give your answer in the required format, do what you can! Marks can be gained at every correct step.

Summary review:

- You need to know and understand all the different symbols for set notation. Not all of them are on page 23 of the *Formulae and Tables* booklet – the ones indicated by * are included in the booklet; you will need to remember the others.

Symbol	Meaning
∈ *	Is an element of
∉	Is not an element of
⊂ *	Is a subset of
⊄	Is not a subset of
=	Is equal to
≠	Is not equal to
#	Cardinal number/number of elements in
∪ *	Union
∩ *	Intersection
𝕌	Universal set
\ *	Set difference
′	Complement
{ } or ∅	Null set

- You need to remember the various different regions of a two set Venn diagram and a three set Venn diagram.
- You need to know how to solve problems involving two sets and three sets.

CHAPTER 3: SETS

Chapter 4: Algebra 1

The topics and skills I will work on are:

Evaluating expressions ☐
Adding, subtracting and simplifying algebraic expressions ☐
Multiplying and simplifying algebraic expressions ☐
Expanding the brackets and simplifying ☐
Simplifying algebraic fractions ☐

Section A: Evaluating Expressions

- To evaluate means to find the value of something. You must substitute the values for the letters into the equations and get a number for an answer. We refer to these letters as variables as the quantity can change or vary, taking on different values. There is no fixed value for what a or x equals.

- The skill of substituting numbers into equations or formulae can be seen throughout maths. You will use this skill not only in Algebra, but also in Arithmetic, Applied Measure, Co-ordinate Geometry, Geometry, Graphing Functions, Patterns and Sequences and Trigonometry to name but a few.

1. If $a = 2$, $b = -1$ and $c = -2$, evaluate each of the following:

Worked example
$2a + 3b - c = 2(2) + 3(-1) - (-2)$
$= 4 - 3 + 2$
$= 3$

Remember!
Always be aware of your signs.
$(+)(+) = +$
$(+)(-) = -$
$(-)(-) = +$
$(-)^2 = +$

a) $2a + 5b - 4$

b) $a^2 - b^3 + c$

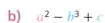

SKILLS FOR EXAM SUCCESS MATHS

c) $\sqrt{a^3 - c^3 - 4b}$. Give your answer in the form $a\sqrt{5}$.

Exam Hint

There is a clue in the question about how to write your answer in order to present it in the required format. If you cannot write the answer in the required format, do as much of the question as you can. Some marks are better than no marks!

2. Plumber A uses the formula $c = 80 + 40h$ to calculate the cost of a job, where c is the cost of the job and h is the number of hours it will take him to complete the work.

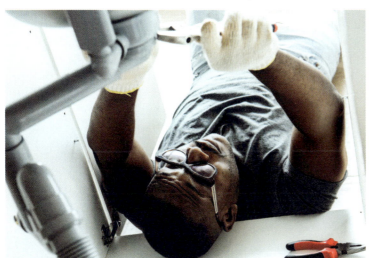

a) Use this formula to calculate the cost of a job that will take 5 hours to complete.

$c = 80 + 40h$

b) Plumber B uses the formula $c = 100 + 35h$. Use this formula to calculate the cost of the same job.

$c = 100 + 35h$

c) Áine calculates that Plumber B is 10% cheaper than Plumber A. Is she correct? Explain your answer.

CHAPTER 4: ALGEBRA 1

Section B: Adding, Subtracting and Simplifying Algebraic Expressions

- You can only add and subtract like terms. You can add or subtract x to or from another x only (for example: $2x + 3x$ or $6x - 3x$), x^2 to or from another x^2 (for example: $2x^2 + 4x^2$ or $5x^2 - 7x^2$) and numbers on their own to or from numbers on their own (for example: $5 + 8$ or $12 - 4$).

- Remember the number line when you are adding and subtracting.

  – When adding numbers, we move right along the number line.

  – When subtracting numbers, we move left along the number line.

- The final step in many algebraic questions involves simplifying the expression and it is important that you are confident in this area. The same rules apply as for adding and subtracting – you can only simplify terms that are the same and x and x^3, for example, are very different terms.

Worked example

Simplify this expression: $\quad 3x^3 + 4x^2 - 3x - 5 - 5x^3 + 2x^2 - 4x + 7$

Like terms are colour-coded here: $\quad 3x^3 + 4x^2 - 3x - 5 - 5x^3 + 2x^2 - 4x + 7$

$$= -2x^3 + 6x^2 - 7x + 2$$

3. Simplify the following expressions.

   a) $7x + 10 + 2x + 5 + x + 3$

   b) $5x^2 - 5x - 10 - 2x^2 - 4x + 8$

   c) $-6x^3 - 9x^2 + 12x - 4x^2 - 6x + 8$

   d) $2 - 8x - 4x^2 - 3x^3 - 3x + 12x^2 - 2x^3$

Section C: Multiplying and Simplifying Algebraic Expressions

Exam Hint
I recommend that you write out each step clearly so that if you do make a mistake, the examiner will be able to pinpoint exactly where your error is and deduct the minimum amount of marks.

Worked example
Multiply out the brackets and simplify.

$= 2(x^2 + 3x - 2) + 4(x^2 - 2x - 3)$

$= 2x^2 + 6x - 4 + 4x^2 - 8x - 12$

$= 6x^2 - 2x - 16$

Step 1: Each term within the brackets must be multiplied by the term outside the brackets.
$2(x^2) = 2x^2 \qquad 2(3x) = 6x \qquad 2(-2) = -4$
$4(x^2) = 4x^2 \qquad 4(-2x) = -8x \qquad 4(-3) = -12$

Step 2: We now simplify our answer.
$2x^2 + 4x^2 = 6x^2 \qquad 6x - 8x = -2x \qquad -4 - 12 = -16$

4. Multiply out the brackets and simplify.

   a) $3(x^3 + 2x^2 - 4x - 3) - 4(2x^3 + 3x^2 - 4x + 1)$

Remember!
Always be aware of your signs. Remember:
$(+)(+) = +$
$(+)(-) = -$
$(-)(-) = +$
$(-)^2 = +$

   b) $-2(-4 + 3x - 2x^2 + 3x^3) + 3(5x - 6 + 3x^3 - 2x^2)$

Worked example
Multiply out the brackets and simplify.

$= x(2x^2 + 3x + 4) - x(x^2 + 3x - 4)$

$= 2x^3 + 3x^2 + 4x - x^3 - 3x^2 + 4x$

$= x^3 + 8x$

Step 1: Each term within the brackets must be multiplied by the term outside the brackets.
$x(2x^2) = 2x^3 \qquad x(3x) = 3x^2 \qquad x(4) = 4x$
$-x(x^2) = -x^3 \qquad -x(3x) = -3x^2 \qquad -x(-4) = 4x$

Step 2: We now simplify our answer.
$2x^3 - x^3 = x^3 \qquad 3x^2 - 3x^2 = 0 \qquad 4x + 4x = 8x$

5. Multiply out the brackets and simplify.

   a) $x^2(x + 3) - 2(x^2 + 4)$

Remember!
$(x)(x) = x^2$
$(x)(x^2) = x^3$

CHAPTER 4: ALGEBRA 1   51

b) $x(3x^2 - 4x - 5) - 2x(-3x^2 + 2x - 1)$

Section D: Expanding the Brackets and Simplifying

Worked example

Remove the brackets and simplify.

$$(x + 3)(x + 4)$$

Step 1: Write the second bracket out twice and split the contents of the first bracket.

$x(x + 4) + 3(x + 4)$

Step 2: Multiply out.

$x^2 + 4x + 3x + 12$

Step 3: Simplify.

$x^2 + 7x + 12$

Two brackets side by side means we must multiply out the brackets. Every term in the first bracket must be multiplied by every term in the second bracket and the final answer given in its simplest form. $(a + b)(x + y)$ $a(x + y) + b(x + y)$

6. Remove the brackets and simplify.

   a) $(2x + 3)(3x - 4)$

   Step 1: Write the second bracket out twice and split the contents of the first bracket.
   $2x(3x - 4) + 3(3x - 4)$

   Step 2: Multiply out.

   Step 3: Simplify.

   b) $(3x - 1)(-4x - 3)$

SKILLS FOR EXAM SUCCESS MATHS

c) $(x + 1)(2x^2 - 3x - 5)$

$x(2x^2 - 3x - 5) + 1(2x^2 - 3x - 5)$

> Do not be put off by the fact that there are three terms in the second bracket here. You follow the exact same technique.
> Step 1: Write the second bracket out twice.
> Step 2: Split the contents of the first bracket.
> Step 3: Multiply out.
> Step 4: Simplify.

d) $(3x - 5)(2x^3 - 4x^2 + 6x - 1)$

e) $(4x - 5)^2$

> Remember: $(4x - 5)^2 = (4x - 5)(4x - 5)$

f) $(3x - 2)^3$

> Remember: $(3x - 2)^3 = (3x - 2)(3x - 2)(3x - 2)$

7. Paul is x years old. Ellie is 5 years younger than Paul.
Sophie is 2 years older than Ellie.
Jack is 4 times the age of Sophie.

Write expressions for:

a) Paul's age _____

b) Ellie's age _____

c) Sophie's age _____

d) Jack's age _____

e) All ages combined _____

CHAPTER 4: ALGEBRA 1   53

Section E: Simplifying Algebraic Fractions

- When adding or subtracting fractions you must first find a common denominator, as covered in Chapter 1.

- You have been adding and subtracting fractions since you were in primary school. The only difference here is that there are a few letters with the numbers too. The technique is exactly the same!

Worked example

Write the following as a single fraction:

$$\frac{x+1}{3} + \frac{x+3}{2}$$

$$= \frac{2(x+1) + 3(x+3)}{6}$$

$$= \frac{2x + 2 + 3x + 9}{6}$$

$$= \frac{5x + 11}{6}$$

Step 1: Find the common denominator. This is done by finding the LCM of 3 and 2.
Step 2: We do to the top what we do to the bottom. We multiplied the 3 in the denominator by 2 so we do the same to the numerator. We multiplied the 2 in the denominator by 3 so we do the same to the numerator.
Step 3: Multiply out the top line.
Step 4: Simplify the top line.

8. Write the following as a single fraction. (Remember, you follow the same steps for subtraction as you do for addition.)

$$\frac{2x+3}{4} - \frac{x-4}{5}$$

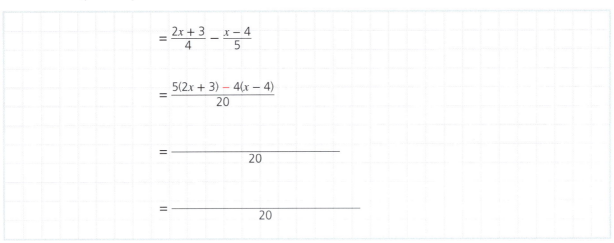

$$= \frac{2x+3}{4} - \frac{x-4}{5}$$

$$= \frac{5(2x+3) - 4(x-4)}{20}$$

$$= \frac{}{20}$$

$$= \frac{}{20}$$

9. Write the following as a single fraction.

$$\frac{3x-5}{3} - \frac{2x+2}{4}$$

Exam Hint

I recommend that you write out each step clearly so that if you do make a mistake, the examiner will be able to pinpoint exactly where your error is and deduct the minimum amount of marks.

Worked example

Write the following as a single fraction. This time, there is algebra in the denominator. We are still going to follow the same technique. We will find a common denominator, work out the top line, multiply out the top line and simplify the top line.

$$= \frac{3}{2x+1} + \frac{2}{3x+2}$$

Step 1: The common denominator is found by multiplying the two denominators together.

$$= \frac{3(3x+2) + 2(2x+1)}{(2x+1)(x+2)}$$

Step 2: We do to the top what we do to the bottom. We multiplied $2x+1$ in the denominator by $3x+2$ so we do the same to the numerator. We multiplied the $3x+2$ in the denominator by $2x+1$ so we do the same to the numerator.

$$= \frac{9x + 6 + 4x + 2}{(2x+1)(3x+2)}$$

Step 3: Multiply out the top line. There is no need to multiply out the bottom line at this stage. This needs to be done only if you are asked to solve for x.

$$= \frac{13x + 8}{(2x+1)(3x+2)}$$

Step 4: Simplify the top line.

---

10. Write the following as a single fraction.

Do not be thrown by the negatives. The − sign is in front of the numerator only. You will follow the exact same technique as before. Remember to be aware of your signs.

$$\frac{5}{2x-1} - \frac{2}{3x-2}$$

11. Write the following expression as a single fraction.

$$\frac{3}{5+4x} - \frac{2}{2x-2}$$

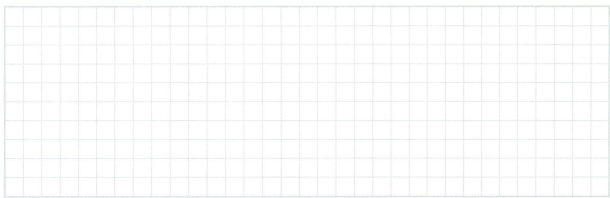

CHAPTER 4: ALGEBRA 1   55

Exam-Style Questions

Some key terms that may appear in algebra questions are:

Evaluate/Work out the value of: Usually when you see this, it involves substituting numbers into an equation and getting a rational number as your final answer.

Simplify: This involves adding and subtracting like terms to get the final answer in its simplest form.

Multiply out and simplify: Here, you will have to multiply out the expression and after you have finished multiplying, simplify your answer by adding or subtracting like terms.

Work out: Show all calculations in the space provided.

Write as a single fraction: You must find a common denominator and add/subtract the fraction.

Verify: You have to prove that the answer you got is correct. This is often done by substituting your answer into an equation that appears in the question.

Exam Hints

1. Attempt every part of every question. You cannot get marks for a blank space!

2. Highlight key words in the question. Make sure you understand what you are being asked.

3. If you are asked to give your answer in a particular format and are unsure of how to do this, do as much as you can.

4. Marks are not just given for the correct answer, they are also available for work completed along the way.

Question 1

Linked Question: Algebra, Numbers

a) If $a = 8$ and $b - a = 5$, work out the value of $4b - 2a$.

Exam Hint

'Work out the value' is another way of asking you to 'evaluate'. Substitute the values for a and b into the equation. Your final answer will be a rational number.

56 SKILLS FOR EXAM SUCCESS MATHS

b) Write the following as a single fraction in its simplest form.

$$\frac{2}{2x+1} - \frac{3}{3x+5}$$

Remember the steps:
1. Find a common denominator (bottom line).
2. Work out your top line.
3. Multiply out the top line.
4. Simplify the top line.
You are not expected to multiply out the bottom line.
SIGNS!

c) The formula to convert degrees Celsius to degrees Fahrenheit is $c = \frac{5}{9}(f - 32)$, where c is the temperature in degrees Celsius and f is the temperature in degrees Fahrenheit.

Find the value of 96 degrees Fahrenheit in degrees Celsius. Give your answer to the nearest degree.

Question 2

a) Find the value of $\dfrac{9a + 6}{\sqrt{28 + a^2}}$ when $a = -6$.

b) Write the following as a single fraction.

$$\frac{5}{2x+1} - \frac{x}{4}$$

CHAPTER 4: ALGEBRA 1   57

Question 3

Linked Question: Algebra, Applied Measure

a) Multiply out and simplify the following.

$(3x - 2)(5 - 4x + x^2)$

> Write the second bracket out twice. Split the contents of the first bracket. Multiply out and simplify.

b) Write the following as a single fraction in its simplest form.

$\dfrac{3}{y-2} - \dfrac{4}{2y+5}$

Exam Hint

Always remember, there are marks for each step in the question, not just the final answer. So even if you are not sure how to complete all of the question, always do as much as you can, no matter how small it might be. You don't know what LPC might be awarded for in the marking scheme.

c) A water bottle is in the shape of a cylinder.

Jane's water bottle has a height of 20 cm and a diameter of 10 cm.

Róisín's water bottle has a height of 25 cm and a diameter of 8 cm.

The formula to calculate the volume of a cylinder is:

$v = \pi r^2 h$

where v is the volume, $\pi = 3.142$, r is the radius and h is the height.

Róisín claims that her bottle is taller and therefore has the greater volume.

Is Róisín correct? Show all your workings.

> You are given the diameter of the bottle in the question and the formula requires the radius.

SKILLS FOR EXAM SUCCESS MATHS

d) You are given two cylindrical flasks.

Both flasks have the same radius (r).

In cylinder 1, the height is 2 times the radius.

In cylinder 2, the height is 3 times the radius.

Find the volume of both cylinders in terms of r.

Leave your answer in terms of π.

All units are in cm.

Remember!
Give your answer in the required format.
Don't forget your units. We are looking at volume here, so units are cubed.

Volume of a cylinder = $\pi r^2 h$

Cylinder 1:   Cylinder 2:

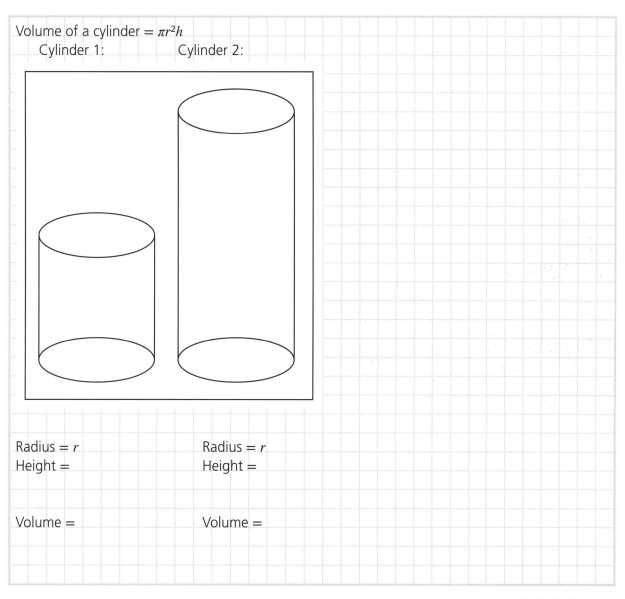

Radius = r Radius = r
Height =   Height =

Volume =   Volume =

Scan this code to access a video with worked solutions for every question.

CHAPTER 4: ALGEBRA 1

Chapter 5: Algebra 2

The topics and skills I will work on are:

Solving equations ☐
Linear equations and linear functions ☐
Solving equations with fractions ☐
Simultaneous equations ☐
Linear inequalities ☐
Changing the subject of a formula ☐
Word equations ☐

Section A: Solving Equations

- When solving equations, the aim is to have anything with a letter on one side of the equals sign and any numbers on their own on the other side.

- The skill of solving equations is used in almost every area of maths and is one of the most important skills you must master.

Worked example

Solve the following equation.

$7x - 6 = 4x + 3$

$7x - 4x = 3 + 6$

$3x = 9$

$x = \frac{9}{3} = 3$

1. Solve the following equations.

   a) $6(2y - 4) = 2(3y - 6)$

   Step 1: Multiply out both sides of the equals sign. Always be aware of your signs.
   Step 2: Balance the equation so that everything with a y is on one side of the equals sign and everything else is on the other side.
   Step 3: Solve for y.

b) $12 + 3(4 - 2a) + 7 = 5(2a - 5) - 2(3a + 7)$

c) Fota Wildlife Park is building a new lion enclosure. One side of the enclosure is to be the shed that the lions sleep in.

The remaining three sides are to be fencing and 96 m of fencing is needed.

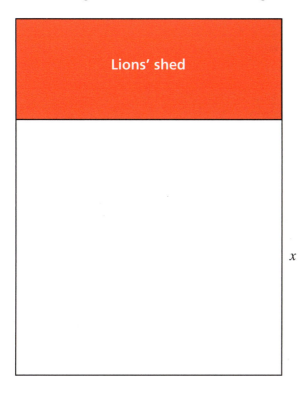

If the width of the fenced enclosure is x:

i. Find, in terms of x, the length of the fencing if the length is 6 m longer than the width.

ii. Use this information to solve for x.

iii. Find the area of the fenced enclosure.　　　Area: Length × Width

CHAPTER 5: ALGEBRA 2　　**61**

Section B: Linear Equations and Linear Functions

- A linear function is a mathematical expression that, when graphed, will form a straight line.
- It is like a linear pattern/sequence. There is a relationship between each term and this relationship can be written as an equation.

2. Simon is training for a marathon. He is currently running 10 km every day. He decides to increase his distance by 0·5 km every day.

   a) Use this information to complete the table showing the distance Simon will run over 7 days.

Day	Distance (km)	Point to be plotted
0	10	(0, 10)
1	10·5	(1, 10·5)
2	11	(2, 11)
3	11·5	
4		
5		
6		
7		

   b) Graph this information. The first few points have been drawn for you. Complete the graph using your data from part (a).

   c) Use this information to form an equation to show how the distance Simon runs changes with time.

   > Simon starts running 10 km on day 0. Every day he increases his distance by 0·5 km.
   > This information can be represented by:
   > Distance = $10 + 0.5d$, where d represents the day number.

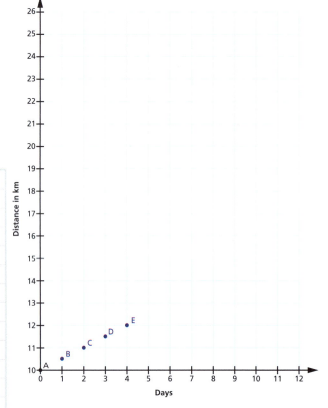

62 SKILLS FOR EXAM SUCCESS MATHS

d) What distance will Simon run in 10 days?

To find the distance run by Simon after 10 days, we substitute 10 for d and evaluate.

$$D = 10 + 0.5d$$

e) How long will it take Simon to be running a distance of 26 km?

To calculate the number of days it will take Simon to run a distance of 26 km, we must put our equation equal to 26 and solve for d.

$$10 + 0.5d = 26$$

Section C: Solving Equations with Fractions

Worked example

Solve the following equation:

$$\frac{2x + 4}{3} - \frac{5x - 7}{2} = 5$$

1. Multiply out both sides of the equals sign. Always be aware of your signs.
2. Balance the equation so that everything with an x is on one side of the equals sign and everything else is on the other side.
3. Solve for x.

$$\frac{2(2x + 4) - 3(5x - 7)}{6} = 5$$

$$\frac{4x + 8 - 15x + 21}{6} = 5$$

$$\frac{-11x + 29}{6} = 5$$

$$\frac{-11x + 29}{6} \diagup \frac{5}{1}$$

$$1(-11x + 29) = 5(6)$$

$$-11x + 29 = 30$$

$$-11x = 30 - 29$$

$$-11x = 1$$

$$x = \frac{1}{-11}$$

Step 1: We need to write the left-hand side as a single fraction. We do this by finding a common denominator, 6, and ensuring what is done to the bottom is also done to the top.
Step 2: Multiply out the top line.
Step 3: Simplify the top line, $4x - 15x = -11x$, $8 + 21 = 29$.
Step 4: To solve for x, we need to form a linear equation.
We place 5 over 1 as all natural numbers are over 1 in fraction form.
We cross multiply: top by bottom $(-11x + 29)1$ and top by bottom $5(6)$.
Cross multiplication is a technique we use to convert an equation in fraction form to an equation in linear form. In order to solve for x, the equation must be in linear form.

CHAPTER 5: ALGEBRA 2

3. Solve the following equation. Give your answer in the form $\frac{m}{n}$ where $m, n \in \mathbb{N}$.

   $$\frac{3x+5}{2} + \frac{x-4}{3} = 16$$

   > Do not be put off by the second part of this question. All this means is you must give your final answer as a fraction. You follow the steps as always.

   In order to be able to solve for x, we will have to cross multiply. However, we cannot cross multiply until the expression on the left is written as a single fraction.

   Step 1: Find the common denominator and the top line.

   $$\frac{()(3x+5) + ()(x-4)}{6} = 16$$

   Step 2: Multiply out the top line. $\dfrac{}{6} = 16$

   Step 3: Simplify the top line. $\dfrac{}{6} = 16$

   Step 4: To solve for x, we need to form a linear equation. Place 16 over 1 as all natural numbers are over 1 in fraction form. Then cross multiply: top by bottom and top by bottom.

   Linear equation:

4. Solve the following equation. Give your answer in the form $\frac{a}{b}$.

   $$\frac{9x-6}{2} = \frac{3x-14}{3} + \frac{9x}{4}$$

   > You have been adding fractions since you were in primary school. Don't let x put you off.

5. The difference between two natural numbers is 6.

   One-third of the larger number minus one-quarter of the smaller number is 5.

   Form an equation in terms of x and solve to find the two numbers.

   > Hint:
   > Smaller number $= x$
   > Larger number $= x + 6$

64 SKILLS FOR EXAM SUCCESS MATHS

Section D: Simultaneous Equations

- It is important to remember when you are solving simultaneous equations that you have to solve for both letters in the equation.

- After you solve for the first letter, make sure that you remember to go back and solve for the second letter. If you don't, it means losing out on precious marks.

Worked example

Solve the simultaneous equations below to find the value of x and the value of y.

$$3x + 2y = 1$$
$$7x + 5y = -2$$

Remember!
Always be aware of your signs.
$(+)(+) = +$
$(+)(-) = -$
$(-)(-) = +$
$(-)^2 = +$

Step 1: We need to isolate x in order to be able to solve for x. To do this we will multiply both equations by different values so that we end up eliminating the ys.

To isolate x, we multiply the top line by 5 and the bottom line by -2. This will set up the simultaneous equations and allow us to eliminate the y values and allow us to solve for x.

$3x + 2y = 1$ × 5 = $15x + 10y = 5$
$7x + 5y = -2$ × −2 = $-14x - 10y = 4$

Step 2: Simplify

$+10y - 10y = 0$
$15x - 14x = x$
$5 + 4 = 9$
$x = 9$

Step 3: We have solved $x = 9$. Now we go back to either of the above equations and solve for y:

$$3x + 2y = 1$$
$$3(9) + 2y = 1$$
$$27 + 2y = 1$$
$$2y = 1 - 27$$
$$2y = -26$$
$$y = -\frac{26}{2}$$
$$y = -13$$

Don't forget to go back and solve for y. With simultaneous equations, you have to get a value for both letters.

Solution: $x = 9$, $y = -13$

CHAPTER 5: ALGEBRA 2   65

6. The equation of the line a is $\frac{x}{2} + \frac{y}{3} - 7 = 0$ and the equation of the line b is $-3x = 10(y - 3)$.

Use simultaneous equations to find the point of intersection of these two lines.

Verify your answer.

Exam Hint
I recommend that you write out each step clearly so that if you do make a mistake, the examiner will be able to pinpoint exactly where your error is and deduct the minimum amount of marks.

Step 1: You must write the equations in a format suitable for solving simultaneous equations.

Equation 1: $\frac{x}{2} + \frac{y}{3} - 7 = 0$

$$\frac{x}{2} + \frac{y}{3} = 7$$

$$(6)\frac{x}{2} + (6)\frac{y}{3} = (6)7$$

$$3x + 2y = 42$$

To get rid of the fractions, you must multiply across the entire equation by the lowest common multiple of 2 and 3.

Equation 2: $\quad -3x = 10(y - 3)$

$$-3x = 10y - 30$$

$$-3x - 10y = -30$$

Multiply out to remove the brackets and rearrange the equation.

Step 2: Solve the simultaneous equations.
$$3x + 2y = 42$$
$$-3x - 10y = -30$$

Step 3: Verify your answer.

66 SKILLS FOR EXAM SUCCESS MATHS

7. Solve the following simultaneous equations:

$\frac{x}{4} - \frac{y}{6} = \frac{5}{6}$

$4x = 3(4 + y)$

Verify your answer.

Exam Hint

Even if you find it difficult to convert the fraction equation into a linear equation, you can still multiply out equation 2. You never know what you might get Low Partial Credit for.

Step 1: Rearrange the equations.

Step 2: Solve the simultaneous equations.

Step 3: Verify your answer.

8. Oliwia and Gabi are make-up artists and are restocking their kits.

Oliwia buys 5 concealers and 4 lipsticks at a cost of €155.

Gabi buys 3 concealers and 7 lipsticks at a cost of €185.

a) Using simultaneous equations, find the cost of a lipstick and a concealer.

Let x = price of a concealer
Let y = price of a lipstick

Applied Arithmetic

b) The bill is inclusive of 23% VAT. Find Oliwia's bill before VAT was added. Give your answer to the nearest euro.

Exam Hint

Always make an effort. Remember there will be marks for every correct step along the way. Some marks are better than no marks! Focus on what you can do within a question.

9. Mahon Point Cinema in Cork hosted a special screening of the new Super Mario movie.

    300 people attended the screening.

    An adult ticket cost €7·50 and a child ticket cost €3·50. The total income is €1550.

    Let x represent the number of adults and y the number of children who attended the cinema.

    Find the number of adults and the number of children that attended the screening.

SKILLS FOR EXAM SUCCESS MATHS

10. $l: 5x + 4y = 24$ and $m: x - 2y = 2$ are two lines. Graph the lines on the same scales and axes and find the point of intersection of these two lines.

Verify your answer using simultaneous equations.

Step 1: In order to graph a line, you need two points on the line. The easiest way to do this is to let $x = 0$ and find the corresponding y-value and then let $y = 0$ and find the corresponding x-value. Each point is written as (x, y) with the x co-ordinate first, followed by the y co-ordinate.

	$5x + 4y = 24$	$x - 2y = 2$
Let $x = 0$	$5(0) + 4y = 24$ $0 + 4y = 24$ $4y = 24$ $\dfrac{4y}{4} = \dfrac{24}{4}$ $y = 6$ Point: $(0, 6)$	$(0) - 2y = 2$ Point:
Let $y = 0$	$5x + 4(0) = 24$ $5x + 0 = 24$ $5x = 24$ $\dfrac{5x}{5} = \dfrac{24}{5}$ $x = 4\cdot8$ Point: $(4\cdot8, 0)$	$x - 2(0) = 2$ Point:

Step 2: Plot the lines. The line l has been plotted for you on the following graph. Now draw the line m and identify the point of intersection of the two lines.

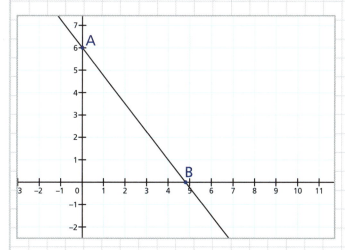

Point of intersection:

Step 3: To verify your answer, you need to prove that your answer from the graph is correct. To do this we use simultaneous equations.

$$5x + 4y = 24$$

$$x - 2y = 2$$

CHAPTER 5: ALGEBRA 2

Section E: Linear Inequalities

Note the differences in the following number lines, especially the differences between plotting natural numbers, integers and real numbers and how to deal with the signs ≤ and <.

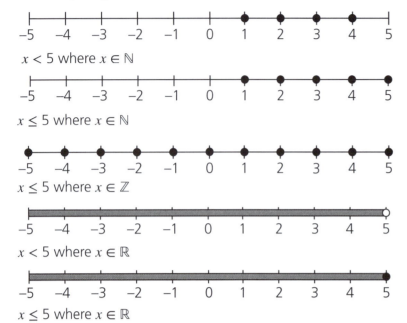

Remember!

Natural numbers (ℕ) are positive whole numbers.

Integers (ℤ) are positive and negative whole numbers.

Real numbers (ℝ) are all numbers, both rational and irrational.

Worked example

Solve the following inequality and plot the solution on the number line.

$6x + 3 \leq 4x + 9$ where $x \in \mathbb{Z}$

$6x + 3 \leq 4x + 9$

$6x - 4x \leq 9 - 3$

$2x \leq 6$

$x \leq 3$

Treat the inequality like you would an equals sign and put everything with a letter to one side and everything without a letter to the other side.

- x is less than or equal to 3.
- It is an integer (positive and negative whole numbers).
- When plotting natural numbers or integers on a number line, we dot the whole numbers.
- We have to plot the integers less than or equal to 3 so we start our plotting at 3 and continue to 0 and below.

11. Find the solution set of $A: 4x + 2 \geq -14$, $x \in \mathbb{Z}$ and $B: 3x - 4 \leq 11$, $x \in \mathbb{Z}$.

    List the elements of the set $A \cap B$.

Remember!

$A \cap B$ = elements found in both the set A and the set B.

When listing the elements of a set we use the following brackets: {}.

70 SKILLS FOR EXAM SUCCESS MATHS

Section F: Changing the Subject of a Formula

- When changing the subject of a formula, the aim is to isolate the required term on one side of the equals sign and move everything else to the other side.

Worked example

Make a the subject of the formula when $s = ut + \frac{1}{2}at^2$.

Our aim here is to have just a on its own on one side of the equals sign and all the other terms on the other side.

$$s = ut + \frac{1}{2}at^2$$

$(2)s = (2)ut + (2)\frac{1}{2}at^2$ | Step 1: Every term is multiplied by 2 to eliminate the fraction. $2s = 2ut + at^2$

$2s = 2ut + at^2$

$2s - 2ut = at^2$ | Step 2: Balance the equation so that everything with an a is on one side and everything without an a is on the other.

$\frac{2s - 2ut}{t^2} = a$ | Step 3: Isolate a.

12. Make u the subject of the formula when $v^2 = u^2 + 2as$.

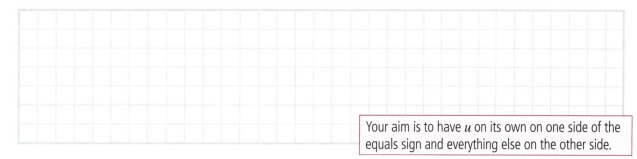

Your aim is to have u on its own on one side of the equals sign and everything else on the other side.

13. a) Make cos A the subject of the formula when $a^2 = b^2 + c^2 - 2bc$ cos A.

Exam Hint

Remember – every step counts. Always make an attempt, however small you feel it is. You never know what LPC will be awarded for in the exam.

b) How would you now make A the subject of the formula?

Exam-Style Questions

Exam Hints

1. Attempt every part of every question. You cannot get marks for a blank space!
2. Highlight key words in the question. Make sure you understand what you are being asked.
3. If you are asked to give your answer in a particular format and are unsure of how to do this, do as much as you can.
4. Marks are not just given for the correct answer, they are also available for work completed along the way.

Question 1

Linked Question: Algebra, Applied Measure

a) Solve the following equation for x.

$$2(3x - 5) + 8 = 4x - 5$$

Remember!
'Solve' means that at the end of the question you need to have $x =$.

Step 1: Multiply out.
Step 2: Move everything with an x to one side of the equals sign and everything else to the other side.

b) Solve for x.

$$\frac{3x + 1}{5} + \frac{x - 2}{2} = \frac{47}{10}$$

Multiply each term by the LCM of the denominators.

SKILLS FOR EXAM SUCCESS MATHS

c) Jodie makes a solid shape using equilateral triangles as faces. It has E edges and F faces, where $E, F \in \mathbb{N}$.

For Jodie's shape:

$\frac{8F}{5} - E = 2$

$3F = 2E$

Solve these simultaneous equations to find the value of E and the value of F.

> Step 1: Rearrange the simultaneous equations so that they read $F, E =$ Number.
> Step 2: Each term in equation 1 must be multiplied by 5 to eliminate the fraction and create a linear equation.
> Step 3: All terms in equation 2 containing letters must be on one side of the equals sign.

d) The formula to find the volume of a cylinder is given by $v = \pi r^2 h$, where $v =$ volume of the cylinder, $r =$ radius of the cylinder and h is the height of the cylinder.

Rearrange the formula to make r the subject of the formula.

Hence, or otherwise, find the radius of the following cylinder, to 1 decimal place.

$v = 108$ cm³
$h = 10$ cm
$\pi = 3.142$

Question 2

Linked Question: Algebra, Numbers, Graphing Functions

a) Describe each of the following sets. Be as specific as possible.

　i.　The set of natural numbers, $\mathbb{N}$.

CHAPTER 5: ALGEBRA 2　　73

ii. The set of integers, $\mathbb{Z}$.

b) Graph the following inequality on the number line below.

$$-3 < x \leq 2, \text{ where } x \in \mathbb{R}$$

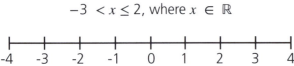

Remember!
When dividing across an inequality by a negative number you must change the direction of the inequality.

c) Use algebra to solve the following inequality.

$$-7 < 8 - 3y \leq 11$$

Question 3

Linked Question: Algebra, Arithmetic, Applied Measure, Numbers, Number Patterns

a) Ellen manages a retail outlet. She has a standard rate cut-off point of €28 320. The standard rate of tax is 20% of income up to the SRCOP and the higher rate of tax is 42% of income above the SRCOP. Her tax payable is €4550 and her tax credits are €4300.

Calculate her gross income for the year.

Gross tax − Tax credits = Tax payable

Let x = amount of money she earns over her standard rate cut off point.

74 SKILLS FOR EXAM SUCCESS MATHS

b) Ellen wants to erect a fence around her house. The fence will go around the two gable ends of the house and the back of the house.

Her house is 20 m long by 13 m wide. She wants the fence to be a distance of x metres from the house.

She purchases 70 m of fencing.

Find, correct to the nearest metre, the distance the fence will be from the house.

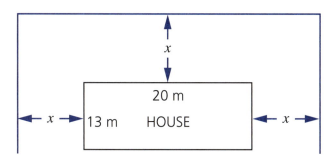

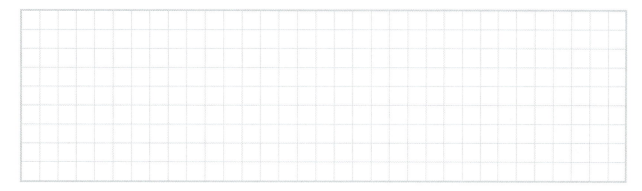

c) Ellen is quoted €4550 as the cost of her fencing. This price is inclusive of 23% VAT.

The warehouse is holding a VAT-free day next month. If Ellen waits, what will she pay for the fencing on the VAT-free day? Give your answer to the nearest euro.

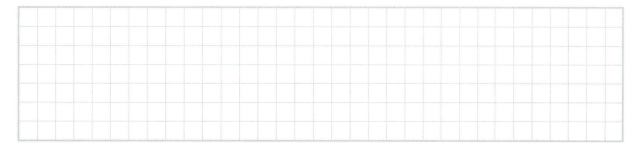

d) Write the cost of the fencing (VAT-free) as a percentage of Ellen's gross income for the year. Give your answer correct to 3 significant figures.

CHAPTER 5: ALGEBRA 2

e) Ellen plants a sunflower in her garden. For the next six months she measures the height of the plant. Her results are as follows:

Month	Height (cm)
March	8
April	11
May	14
June	17
July	20
August	23

i. Using this information, form an expression to illustrate the growth of the plant.

ii. Hence or otherwise, find the height of the plant after 12 months.

iii. Is this an example of a linear function? Justify your answer.

Part (iii) is a two-part question. There would be marks available for stating whether the data is an example of a linear function **and** for justifying your answer.

Scan this code to access a video with worked solutions for every question.

76 SKILLS FOR EXAM SUCCESS MATHS

Chapter 6: Algebra 3

The topics and skills I will work on are:

Factors – Highest common factor
Factors by grouping
Factors – Difference of two squares
Factorising quadratic trinomials
Factorising more complex quadratic trinomials
Solving equations using factors
Simplifying expressions using factors
Solving quadratic equations using the quadratic formula $x = \dfrac{-b \pm \sqrt{b^2 - 4ac}}{2a}$
Long division in algebra
Word equations

Section A: Factors – Highest Common Factor

- When asked to factorise, you are looking for the terms that are multiplied together to get the given answer.
- Factorising is the reverse of expanding the brackets.

1. Factorise the following.

   a) $5x^2 + 10x$
      $5x (\quad)$

   b) $3ab + 6ab^2 - 9a^2b$
      $(\quad)$

   c) $-2x^3 + 4x^2 - 6x$
      $(\quad)$

   d) $xy - 5x^2y + 6xy^2$
      $(\quad)$

   Step 1: Look at each term in the expression.
   Step 2: Put the HCF outside the bracket.
   Step 3: Divide each term by the HCF to get what goes inside the bracket.

Section B: Factors by Grouping

Worked example

Factorise the following.

$ax + ay + cx + c$

$\boxed{ax + ay} + cx + cy$

$a(x + y) + \boxed{cx + cy}$ Step 1: Look at the first pair of terms, put the HCF outside the bracket, fill in inside the bracket.

$a(x + y) + c(x + y)$ Step 2: Look at the second pair of terms, put the HCF outside the bracket, fill in inside the bracket.

$(a + c)(x + y)$ Step 3: Take the two HCFs and put them together in one pair of backets, next to each other. The expression is now factorised.

2. Factorise the following.

a) $a^2 - ab + ac - bc$

b) $7p - 3qx - 7q + 3px$

c) $x^2 - 6y + 3x - 2xy$

> Sometimes you will have to rearrange the expression before you start to factorise. Addition is commutative which means the order of the terms does not matter.

Section C: Factors – Difference of Two Squares $(a-b)(a+b)$

- To factorise an expression such as $a^2 - b^2$, you must:
  - have a – sign between both terms **and**
  - write each term as a perfect square with brackets.
- Remember, 1 is a square number.

Worked example

Factorise the following.

$= x^2 - 25$

$a b$

$= (x)^2 - (5)^2$

$(a - b)(a + b)$

$= (x - 5)(x + 5)$

78   SKILLS FOR EXAM SUCCESS MATHS

3. Factorise the following.

   a) $100y^2 - 225x^2$

   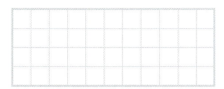

   b) $y^2 - 1$

   c) $9a^2 - 36b^2$

   d) $49 - p^2$

Section D: Factorising Quadratic Trinomials

4. Factorise the following.

   a) $x^2 + x - 6$
   $(x \quad)(x \quad)$ Step 1: Start with two pairs of brackets. The x^2 term comes from multiplying x by x.

   $(x - 2)(x + 3)$ Step 2: You need two numbers that **multiply** to give -6 and the same two numbers must **add** to give $+1$ (the coefficient of x).

   b) $x^2 - 11x + 28$

   c) $x^2 - 2x - 48$

   d) $x^2 + 20x + 91$

Section E: Factorising More Complex Quadratic Trinomials

Worked example

Factorise the following.

$$2x^2 + 11x + 12$$

Here the coefficient in front of the x is greater than 1 so I will use a different technique to the one used in question 4. Either technique is perfectly acceptable in the exam.

Step 1: Take the number in front of the x^2 and the last term and multiply them together.

$$2 \times 12 = 24$$

Step 2: We now need two numbers that multiply to give 24 and the same two numbers must add together to give 11 (the number in front of the x). List the factors of 24: 1×24, 2×12, 3×8, 4×6.

To meet the requirements of multiplying to give 24 and adding to give 11, +8 and +3 are the factors we need.

Step 3: We now have to rewrite the original equation. Instead of $11x$, we substitute in our factors.

$$2x^2 + 11x + 12$$

$$2x^2 + 8x + 3x + 12$$

Look at the first pair of terms, put the HCF outside the bracket, fill in inside the bracket.

Step 4: Factorise by grouping.

$$2x^2 + 8x \ + \ 3x + 12$$

Look at the second pair of terms, put the HCF outside the bracket, fill in inside the bracket.

$$2x(x + 4) + 3(x + 4)$$

Take the terms outside the brackets and pair them together in one set of brackets, followed by what is already written in the brackets.

$$(2x + 3)(x + 4)$$

5. Factorise the following.

    a) $6x^2 - 13x - 28$

    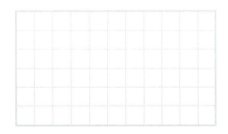

    b) $15x^2 - 7x - 2$

    c) $12x^2 - 29x + 15$

    Exam Hint

    I recommend that you write out each step clearly so that if you do make a mistake, the examiner will be able to pinpoint exactly where your error is and deduct the minimum amount of marks.

SKILLS FOR EXAM SUCCESS MATHS

Section F: Solving Equations Using Factors

- In graphing functions, solving the quadratic equation tells us where the curve will cut the x-axis, i.e. the roots of the equation.
- To solve means to get the answer using algebra/numerical or graphical methods.

Worked example

Find the roots of the following equation: $2x^2 + 7x + 3 = 0$.

Step 1: Take the coefficient, the number in front of the x^2, and the last term and multiply them together.

$2 \times 3 = 6$

Step 2: We now need two numbers that multiply to give 6 and the same two numbers must add together to give 7 (the number in front of the x). We list the factor pairs of 6: 1×6, 2×3.

The two numbers are $+6$ and $+1$.

Step 3: We now have to rewrite the original equation. Instead of $7x$, we substitute in our factors.

$2x^2 + 6x + x + 3 = 0$

$2x(x + 3) + 1(x + 3) = 0$

$(2x + 1)(x + 3) = 0$

Look at the first pair of terms, put the HCF outside the bracket, fill in inside the bracket.
Look at the second pair of terms, put HCF outside the bracket, fill in inside the bracket.

Step 4: We now put each factor equal to 0 and solve for x.

$2x + 1 = 0 \qquad x + 3 = 0$

$2x = -1 \qquad x = -3$

$x = \dfrac{-1}{2}$

If you are asked to **factorise an equation**, you stop at Step 3. If you are asked to **solve for** x, you continue and complete Step 4.

6. Solve for x in the following equations.

   a) $4x^2 + 9x - 9 = 0$

b) $3x^2 - 5x - 12 = 0$

c) $5x^2 + 3x - 2 = 0$

Section G: Simplifying Expressions Using Factors

7. Use factors to simplify the following.

$$\frac{2f^2 + f - 15}{f^2 - 9}$$

Step 1: Factorise the numerator. We practised this type of question in the last section.
Step 2: Factorise the denominator using the difference of two squares. Take it one step at a time. You can use either type of factorising. Do not get overwhelmed when you see a couple of different types of equation within one question.
Step 3: Simplify your answer.

8. Use factors to simplify the following.

$$\frac{2x^2 + 4x}{2x^2 + x - 6}$$

Hint:
Numerator: HCF
Denominator: Quadratic trinomial

Section H: Solving Quadratic Equations Using the Quadratic Formula

- When asked to solve a quadratic equation (get the answer using algebra/numerical or graphical methods), there are two ways you can approach it:
  - Factorise the equation, put each factor equal to 0 and solve for x.
  - Use the formula $x = \frac{-b \pm \sqrt{b^2 - 4ac}}{2a}$. This is provided in the *Formulae and Tables* booklet (p. 20).
- Oftentimes, if you are under pressure in an exam, it is less time consuming to use the $-b$ formula to solve for x.

Worked example

Solve the equation $2x^2 - 7x - 3 = 0$ and give your answer correct to 2 decimal places.

Step 1: Identify a, b and c.

$a = 2 \qquad b = -7 \qquad c = -3$

Step 2: Substitute values into your formula.

$$x = \frac{-b \pm \sqrt{b^2 - 4ac}}{2a}$$

$$x = \frac{-(-7) \pm \sqrt{(-7)^2 - 4(2)(-3)}}{2(2)}$$

$$x = \frac{7 \pm \sqrt{73}}{4}$$

$x = \frac{7 + \sqrt{73}}{4}$ or $x = \frac{7 - \sqrt{73}}{4}$

$x = 3 \cdot 886 \qquad\qquad x = -0 \cdot 386$

$x = 3 \cdot 89 \qquad\qquad x = -0 \cdot 39$

> If you are asked to give your answer to a particular number of decimal places or are asked to give your answer in the form $a \pm \sqrt{b}$, this is a hint that you must use the $-b$ formula.

> If you end up with a negative number in the square root you have gone wrong somewhere.

> Don't forget to round to 2 decimal places for the last few precious marks.

Exam Hint

I recommend writing out the values for a, b and c so that they are clear to both you and the examiner.

Always write out the formula you are using. The correct formula with any correct, relevant substitution merits LPC in an exam.

CHAPTER 6: ALGEBRA 3

9. Solve the equation $x^2 - 6x - 23 = 0$.

   Give your answer in the form $a \pm b\sqrt{2}$, where $a, b \in \mathbb{Z}$.

Exam Hint

Do not be put off if you are asked to give your answer in this format. Do as much as you can – remember, there are marks for each step along the way.

Write out each step clearly so that if you do make a mistake, the examiner will be able to pinpoint exactly where your error is and deduct the minimum amount of marks.

Hint:
Numerator: HCF
Denominator: Quadratic trinomial

Worked example

The product of two consecutive odd natural numbers is 195. One of the numbers is x.

Use this information to form a quadratic equation and hence, or otherwise, solve for x.

First number $= x$

Next odd number $= x + 2$

'Product' means multiply so:

$$x(x + 2) = 195$$

$$x^2 + 2x = 195 \quad \text{Multiply out.}$$

$$x^2 + 2x - 195 = 0$$

To factorise, we need two numbers that multiply together to give -195 and the same two numbers must add together to give $+2$. We list the factor pairs of 195 to help figure out which numbers: 1×195, 3×65, 5×39, 13×15.

$(x - 13)(x + 15) = 0$

$\quad x - 13 = 0 \quad x + 15 = 0$

$\quad\quad x = 13 \quad\quad x = -15$

You can use either technique to factorise this quadratic trinomial. I chose to 'open the brackets' rather than use the $-b$ formula so that you have an example of both.

So $x = 13$ as we are told in the question that the numbers are natural numbers.

$\quad x + 2 = 15$

It is important to identify which value for x you feel is correct to show full understanding of the question.

84 SKILLS FOR EXAM SUCCESS MATHS

10. Find three consecutive natural numbers such that the product of the first two is equal to 4 times the third number minus 4.

    Hint:
    First number $= x$
    Second number $= x + 1$
    Third number $= x + 2$

11. In the diagram below, the length of each of the sides is given in terms of x, where $x \in \mathbb{N}$.

    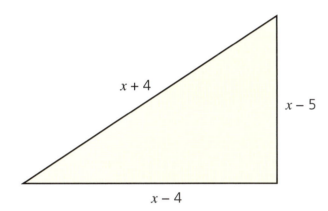

    Show that there is **only one** value of x for which the triangle is right angled.

    Theorem of Pythagoras: $c^2 = a^2 + b^2$

Section I: Long Division in Algebra

- The technique shown in the following worked example can also be applied when working with graphing functions.
- An alternative method is to factorise the trinomial first and set the question up in fraction form. Either technique is perfectly acceptable in the exam.
- Sometimes, you can be given a quadratic or a cubic equation and one of its factors.
- You could be asked to use this information to find the other root/roots of the equation.

Worked example

Divide $5x^2 + 27x + 28$ by $x + 4$.

$$
\begin{array}{r}
5x \\
x+4 \overline{\smash{\big)}\, 5x^2 + 27x + 28} \\
5x^2 + 20x
\end{array}
$$

$5x^2$ divided by $x = 5x$
Place $5x$ above the line.
Multiply $5x(x+4)$ to get the second line: $5x^2 + 20x$

$$
\begin{array}{r}
5x \\
x+4 \overline{\smash{\big)}\, 5x^2 + 27x + 28} \\
-5x^2 - 20x \\
\overline{ 7x + 28}
\end{array}
$$

Change the signs of the bottom line.
$5x^2 + 20x$ becomes $-5x^2 - 20x$
$5x^2 - 5x^2 = 0$, $27x - 20x = 7x$, bring down the 28.

$$
\begin{array}{r}
5x + 7 \\
x+4 \overline{\smash{\big)}\, 5x^2 + 27x + 28} \\
-5x^2 - 20x \\
7x + 28 \\
7x + 28
\end{array}
$$

$7x$ divided by $x = 7$
Place 7 above the line
Multiply $7(x + 4)$ to get the second line $7x + 28$

$$
\begin{array}{r}
5x + 7 \\
x+4 \overline{\smash{\big)}\, 5x^2 + 27x + 28} \\
-5x^2 - 20x \\
7x + 28 \\
-7x - 28 \\
\overline{ 0}
\end{array}
$$

Change the signs of the bottom line so that $7x + 28$ becomes $-7x - 28$.
$7x - 7x = 0$, $+28 - 28 = 0$
If you do not get 0 here at this final stage, you have made a mistake.

---

12. Divide $6x^3 + 24x^2 + 34x + 20$ by $2x + 4$.

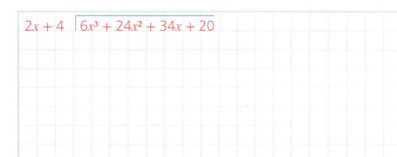

Exam Hint

This is the minimum that you should do. Sometimes LPC is awarded for setting up the division like this.

13. Show that $x - 2$ is a factor of $8x^3 - 18x^2 - 11x + 30$.

    Hence, or otherwise, find the other two factors.

1. Carry out long division.
2. When you divide $8x^3 - 18x^2 - 11x + 30$ by $x - 2$, you will get a quadratic equation. You must factor this quadratic equation to get the other two factors of this cubic equation.

86   SKILLS FOR EXAM SUCCESS MATHS

Exam-Style Questions

Exam Hints

1. Attempt every part of every question. You cannot get marks for a blank space!
2. Highlight key words in the question. Make sure you understand what you are being asked.
3. Write the relevant formula from the *Formulae and Tables* booklet. The correct formula with some relevant substitution will get you marks.
4. If you are asked to give your answer in a particular format and are unsure of how to do this, do as much as you can.
5. Marks are not just given for the correct answer, they are also available for work completed along the way.

Question 1 – 2022

Linked Question: Algebra, Geometry, Number Patterns

The three triangles A, B and C are shown below.

The given lengths of the sides of each triangle are in centimetres, where $x, y \in \mathbb{N}$.

In this question, take 'the perimeter' to mean 'the length of the perimeter'.

a) The perimeter of Triangle A is 8 cm.

   Two of the sides have length 2 cm and 3·5 cm, respectively, as shown.

   Work out the length of the third side of Triangle A.

   Remember! Perimeter is all three sides added together.

CHAPTER 6: ALGEBRA 3 | 87

b) i. Write down the perimeter of Triangle B, in terms of x.

ii. The perimeter of Triangle B is 24 cm.

Use this to work out the value of x.

c) The perimeters of the three triangles A, B and C form a linear sequence.

Triangle C has the largest perimeter.

i. The perimeter of Triangle C is k cm, where $k \in \mathbb{N}$. Find the value of k.

Hint: In a linear sequence, the difference between each term is constant.

ii. Hence work out the value of y, where $y \in \mathbb{N}$.

SKILLS FOR EXAM SUCCESS MATHS

Question 2

a) Factorise fully $5fh - 2h^2 - 6h + 15f$.

Hint: Factorise by grouping.

b) Solve the equation $2x^2 + 2x - 1 = 0$.

Give your answer in the form $\frac{-a + a\sqrt{3}}{b}$.

There is a hint in the question as to how to solve this problem. You are asked to give your answer in surd form, therefore you use the $-b$ formula.

c) Simplify the following.

$\frac{3x^2 + 22x + 35}{x^2 - 25}$

Numerator – factorising a quadratic trinomial
Denominator – difference of two squares

d) Divide $10x^3 - 26x^2 + 16x - 8$ by $2x - 4$.

Exam Hint
The minimum you should do here is set up your division. Every step counts!

CHAPTER 6: ALGEBRA 3

Question 3

Linked Question: Algebra, Applied Measure, Arithmetic, Trigonometry, Numbers

The local GAA club wants to convert some ground into a training pitch with a walking track around it. The ground is 136 m long and 86 m wide. The pitch will be located in the centre of the ground with the track all around the outside.

Exam Hint
I recommend that you write out each step clearly so that if you do make a mistake, the examiner will be able to pinpoint exactly where your error is and deduct the minimum amount of marks.

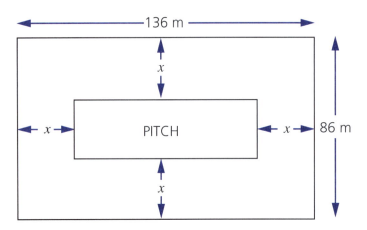

a) If the area of the pitch is 10 400 m² and the track is to be a consistent width on all four sides of the pitch, how wide will the track be?

Hint: Length of pitch = $136 - x - x$

b) Hence, or otherwise, find the total area of the track.

Don't forget your units!

c) For every 2 m² of tarmac, the company will charge €150.

How much will it cost to tarmac the track?

d) ECON, the local supplier of solar lights, agrees to sell the club 30 lights VAT-free.
The VAT rate is 18·5%.

Inclusive of VAT, each light costs €249.

What will the total cost of the lights be to the club, to the nearest euro?

e) During a training session, Rory forms part of the 'wall' in front of Piotar as he takes a free kick.

Rory is 1·82 m tall and is a distance of 8·5 m from the ball.

Calculate the angle of elevation, x, from the ball to the top of Rory's head. Give your answer correct to the nearest whole number.

Question 4

Linked Question: Algebra and Graphing Functions

a) The graphs of the functions $f(x) = x^2 + 2x - 3$ and $g(x) = -x^2 - 2x + 3$ are shown below.
Identify each graph by writing $f(x)$ or $g(x)$ in the space provided below each graph.

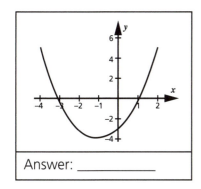

Answer: _____

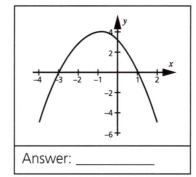

Answer: _____

CHAPTER 6: ALGEBRA 3

b) The graphs of the functions $y = h(x)$ and $y = k(x)$ are shown below.

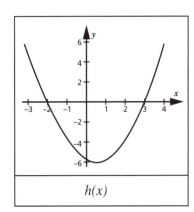
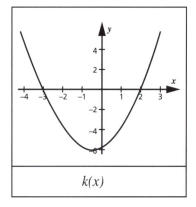

$h(x)$ $k(x)$

Write down the roots of each function.

Hence, or otherwise, write down the equation for each function.

Roots of $h(x)$ =
Equation of $h(x)$ =

Roots of $k(x)$ =
Equation of $k(x)$ =

Summary review:

- You do not need to remember the $-b$ formula. It is in the *Formulae and Tables* booklet.

- If asked to factorise, check to see which of the four techniques you will be using: HCF, factors by grouping, quadratic trinomials or difference of squares.

- If you are asked to solve for x, you need to find a value for x as your solution.

- When asked to solve for x, using the $-b$ formula is often the easiest way to get the solution.

Scan this code to access a video with worked solutions for every question.

SKILLS FOR EXAM SUCCESS MATHS

Chapter 7: Co-ordinate Geometry of the Line

The topics and skills I will work on are:

Co-ordinate plane, distance and midpoint
- Plotting points on the co-ordinate plane
- Using the midpoint and distance formulae

Slope and equation of a line
- Finding the slope and equation of a line
- Giving the equation of a line in the form $ax + by + c = 0$ and $y = mx + c$

Graphing lines
- Interpreting data about slope and y-intercept from the equation of a line formula

Parallel and perpendicular lines
- Determining whether lines are parallel or perpendicular by examining the slopes
- Finding the equation of a line that is parallel/perpendicular to a given line

Finding the point of intersection of two lines by a) graphing lines and b) using simultaneous equations.

Section A: Co-ordinate Plane, Distance and Midpoint

1. Plot the following points on the co-ordinate plane below:

   $A = (4, 5)$ $B = (-2, 3)$ $C = (1, -4)$ $D = (-6, -5)$ $E = (0, 5)$ $F = (-5, 0)$

   The point A has been plotted for you.

   Remember!
   When plotting points, you go to the x-axis first and then the y-axis.

   For the point A (4, 5), you go to 4 on the x-axis and 5 on the y-axis and plot.

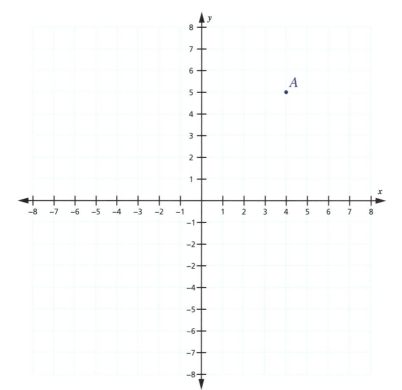

CHAPTER 7: CO-ORDINATE GEOMETRY OF THE LINE 93

Worked example

Using the points *B* and *D* from question 1, calculate the midpoint of [*BD*].

Midpoint: $\frac{x_1 + x_2}{2}, \frac{y_1 + y_2}{2}$ $\quad x_1 = -2$

$= (\frac{-2-6}{2}, \frac{3-5}{2})$ $\quad y_1 = 3$

$= (\frac{-8}{2}, \frac{-2}{2})$ $\quad x_2 = -6$

$= (-4, -1)$ $\quad y_2 = -5$

Exam Hint

Always write the formula from the *Formulae and Tables* booklet (p. 18) into your answer booklet. The correct formula with any relevant substitution will merit marks in the exam.

2. *C* is the midpoint of the line segment [*DG*]. Calculate the co-ordinates of the point *G*.

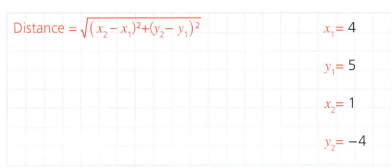

Add 7 to the *x* co-ordinate and add 1 to the *y* co-ordinate.

3. Calculate the length of the line segment [*AC*]. Give your answer in the form $a\sqrt{b}$.

Distance $= \sqrt{(x_2 - x_1)^2 + (y_2 - y_1)^2}$

$x_1 = 4$
$y_1 = 5$
$x_2 = 1$
$y_2 = -4$

Remember!

Always be aware of your signs.
$(+)(+) = +$
$(+)(-) = -$
$(-)(-) = +$
$(-)^2 = +$

Section B: Slope and Equation of a Line ($ax + by + c = 0$ and $y = mx + c$)

4. Find the slopes of the lines *AB* and *CD*.

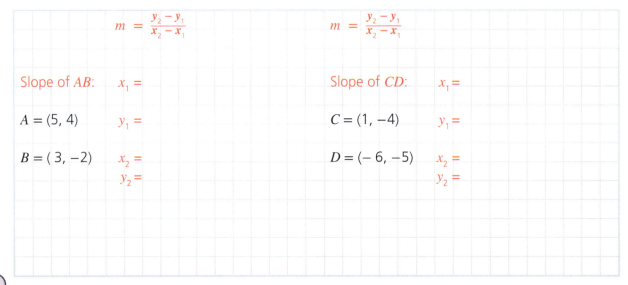

94 SKILLS FOR EXAM SUCCESS MATHS

> **Remember!**
> To find the equation of a line you need:
> 1. a point on the line (x_1, y_1)
> 2. the slope of the line (m).
>
> In the equation of the line, we sub in values for (x_1, y_1) and m but nothing for x and y.

5. Find the equation of the line AB. Give your answer in the form $ax + by + c = 0$.

   Step 1: Slope $(m) = \dfrac{y_2 - y_1}{x_2 - x_1}$ or $\dfrac{\text{Rise}}{\text{Run}}$

   $m =$

   $x_1 = 5$
   $y_1 = 4$
   $x_2 = 3$
   $y_2 = -2$

   Step 2: Equation of a line: $y - y_1 = m(x - x_1)$

   > **Exam Hint**
   > If you are unsure how to give your answer in a particular format, do not let this stop you attempting the question! You will earn a significant portion of the marks for the work you have done before this final step.

   Worked example
   Find the equation of a line containing the point $(-2, 3)$ with a slope $m = -\dfrac{4}{5}$

   How to deal with finding the equation of a line when the slope is a fraction

$y - y_1 = m(x - x_1)$	
$y - 3 = -\dfrac{4}{5}(x - -2)$	Sub the values into the formula.
$y - 3 = -\dfrac{4}{5}(x + 2)$	Sort out the signs (remember $(-)(-) = +$).
$5(y - 3) = -4(x + 2)$	Convert to an equation without any fraction.
$5y - 15 = -4x - 8$	Multiply out the equation.
$4x + 8 + 5y - 15 = 0$	Bring all the terms to one side of the equals sign.
$4x + 5y - 7 = 0$	Simplify and give the answer in the form $ax + by + c = 0$.

   $x_1 = -2$
   $y_1 = 3$
   $m = -\dfrac{4}{5}$

6. The point E is $(1, 4)$ and F is $(2, 5)$. Investigate if these points are on the line AB.

   > **Remember!**
   > If the left-hand side (LHS) = the right-hand side (RHS), the point is on the line.
   > If LHS ≠ RHS, the point is not on the line.
   > Explain your answer so that the examiner knows you understand.

CHAPTER 7: CO-ORDINATE GEOMETRY OF THE LINE

Worked example

The equation of a line is:

$$2x + 3y - 9 = 0$$

a) Write the equation of the line in the form $y = mx + c$.

b) Hence, find the slope of the line (m) and where the line intercepts the y-axis.

a) $2x + 3y - 9 = 0$

$3y = -2x + 9$ — Leave y on one side of the equation and move all other terms across the equal sign.

$y = -\frac{2}{3}x + \frac{9}{3}$ — We need a single y as the formula states $y = mx + c$.

$y = -\frac{2}{3}x + 3$

$y = mx + c$

b) Comparing our rearranged equation with $y = mx + c$, we can see that:

- the slope (m) is $-\frac{2}{3}$
- the line intercepts the y-axis at (0, 3).

7. The equation of a line is:

$$3x - 2y - 12 = 0$$

a) Write the equation of the line in the form $y = mx + c$.

b) Hence, find the slope of the line (m) and the y-intercept.

a) Rearrange the equation: $3x - 2y - 12 = 0$

b) Compare the rearranged equation with $y = mx + c$

SKILLS FOR EXAM SUCCESS MATHS

Section C: Graphing lines

Remember!

- You only need two points in order to be able to draw a straight line.
- All the way along the x-axis, $y = 0$.
- By letting $y = 0$, we can find the point on the x-axis that is on the line.
- All the way along the y-axis, $x = 0$.
- By letting $x = 0$, we can find the point on the y-axis that is on the line.

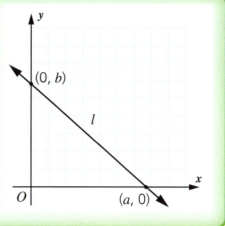

8. The lines a and b are on the co-ordinate plane:

   $a: x - y = 1$ $b: 2x - 3y = 6$

Worked example

a) Graph the lines and use your graph to find the point of intersection.

$$a: x - y = 1 \qquad b: 2x - 3y = 6$$

Step 1: Let $x = 0$, $0 - y = 1$ $2(0) - 3y = 6$

Find y $-y = 1$ $-3y = 6$

                        $y = -1$ $y = \frac{6}{-3} = -2$

                        Point: $(0, -1)$ Point: $(0, -2)$

Step 2: Let $y = 0$, $a: x - y = 1$ $b: 2x - 3y = 6$

Find x $x - 0 = 1$ $2x - 3(0) = 6$

                        $x = 1$ $2x = 6$

                                           $x = \frac{6}{2} = 3$

                        Point: $(1, 0)$ Point: $(3, 0)$

Remember!

Plot the points (x first, then y) and draw the lines on the co-ordinate plane. State the point of intersection of the two lines.

If the lines do not intersect, extend them until they do.

CHAPTER 7: CO-ORDINATE GEOMETRY OF THE LINE

The line *a* contains the points (0, −1) and (1, 0)

The line *b* contains the points (0, −2) and (3, 0)

Point of intersection: _____

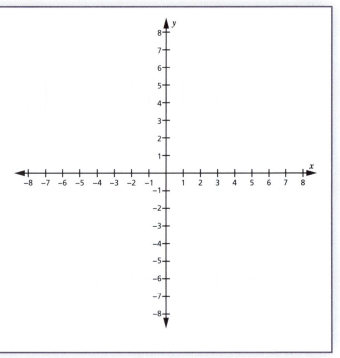

Worked example

b) Verify, using algebra, that your answer is correct.

Step 1: $x - y = 1$ — In order to find the value of x, we have to eliminate y.

$2x - 3y = 6$

Step 2: $3x - 3y = 3$ — To achieve this, every term in the top line is multiplied by 3
and
We multiply every term in the bottom line by −1.

$2x - 3y = 6$

Step 3: $3x - 3y = 3$ — Add the two equations to eliminate y.

$-2x + 3y = -6$

$x = -3$

Step 4: $x - y = 1$ — Return to your top equation. Sub in your value for x. Find y.

$-3 - y = 1$

$-3 - 1 = y$

$-4 = y$

Point of intersection: (−3, −4)

98 SKILLS FOR EXAM SUCCESS MATHS

9. Two people are taking part in two different charity walks. One is making his way from Sligo to Cork. His route can be represented by the equation 2x + 3y = 15. The other is making her way from Limerick to Waterford. Her route can be represented by the equation 5x + y = −8.

   a) Find the point where their routes overlap by graphing the two lines on the co-ordinate plane.

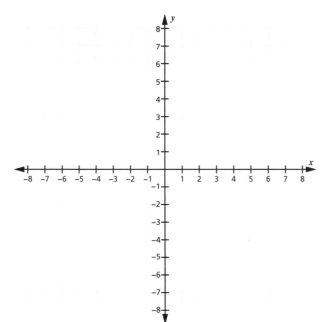

   2x + 3y = 15

   Let x = 0. Find y

   Let y = 0. Find x

   Two points on this line are:
   (_ , _) and (_ , _)

   5x + y = −8

   Two points on this line are:
   (_ , _) and (_ , _)

   b) Use algebra to verify your answer.

   Solve the equations simultaneously:
   2x + 3y = 15
   5x + y = −8

   It's OK to have fractions in your answers for the points in this question.

   c) Which method do you think gives the more accurate answer?

   Explain your reasoning.

Section D: Parallel and Perpendicular Lines

- Parallel lines have the same slope.
- To find a perpendicular slope, 'invert the slope and change the sign'.

Example:

Slope of a line	Slope of a parallel line	Slope of a perpendicular line
$\frac{2}{3}$	$\frac{2}{3}$	$-\frac{3}{2}$
-4	-4	$\frac{1}{4}$
$\frac{5}{6}$	$\frac{5}{6}$	$-\frac{6}{5}$

Worked example

Complete the table below to show the equation of each line in the diagram.

Line	Equation
r	$y = 2x - 4$
n	$y = x$
s	$y = -x$
p	$y = 2x + 4$

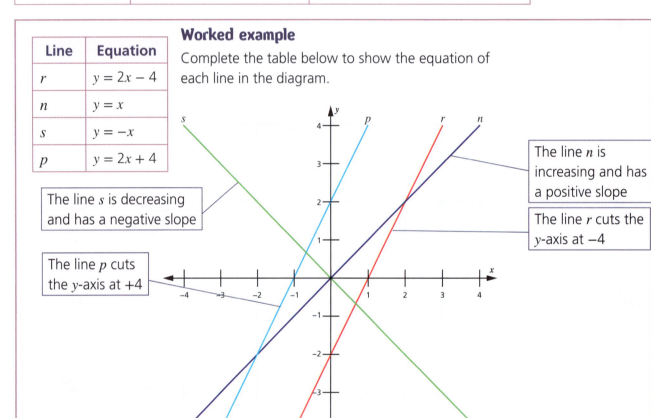

The line s is decreasing and has a negative slope

The line p cuts the y-axis at $+4$

The line n is increasing and has a positive slope

The line r cuts the y-axis at -4

Remember!

If you rearrange the equation of a line into the form $y = mx + c$, then the number in front of the x is the slope (m) of the line and the final term (c) is the y-intercept.

Example:

$3x + 4y - 5 = 0$

$4y = -3x + 5$

$y = -\frac{3}{4}x + \frac{5}{4}$

$y = mx + c$

Slope (m) $= -\frac{3}{4}$, y-intercept $= (0, \frac{5}{4})$

All along the length of the y-axis, $x = 0$, so your point will be in the form $(0, \frac{5}{4})$.

10. Complete the table below to show the equation of each line on the diagram.

    q is parallel to s and r is parallel to t.

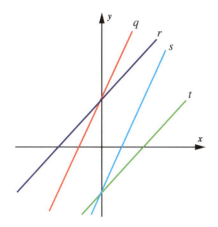

Equation	y-intercept	Slope	Parallel to another line? (same slope)	Perpendicular to another line? ($m_1 \times m_2 = -1$)	Line
$y = x + 3$					
$y = x - 3$					
$y = 2x + 3$					
					s

Worked example

The lines $p: 5x + 6y - 43 = 0$ and $q: 6x - 5y + 21 = 0$ are two lines on the co-ordinate plane.

Prove that they are perpendicular to each other.

To prove if two lines are perpendicular or parallel to each other, we first need to find the slopes of both lines.

There are two methods that we mainly use to identify the slopes of a line from their equation.

Method 1 has been used to identify the slope of the line p and method 2 to find the slope of the line q.

$p: 5x + 6y - 43 = 0$

$5x + 6y - 43 = 0$

$6y = -5x + 43$

$y = -\frac{5}{6}x + \frac{43}{6}$

$m = \frac{-5}{6}$

Rearrange the equation into the form $y = mx + c$, where y is on its own on one side of the equals sign and everything else is in the other side.

$q: 6x - 5y + 21 = 0$

$6x - 5y + 21 = 0$

Slope $= \frac{-6}{-5}$

Slope $= \frac{6}{5}$

$\frac{-5}{6} \times \frac{6}{5} = -1$

Remember!

Slope $(m) = -\frac{a}{b}$ where

a = the sign and number in front of the x

b = the sign and number in front of the y

Conclusion: $m_1 \times m_2 = -1$, this shows that the lines are perpendicular to each other.

Exam Hint

It is very important that you include this line in your answer to show the examiner that you understand what the answer means and to get all the marks available for this question.

CHAPTER 7: CO-ORDINATE GEOMETRY OF THE LINE

11. The lines $p: -2x + y + 5 = 0$ and $k: x + 2y - 23 = 0$ are on the co-ordinate plane.

    a) Investigate if they are perpendicular to each other.

    b) Find where the line p cuts the y-axis.

    c) Find where the line k cuts the y-axis.

    a) Step 1: Slope of the line p (m_1)

    Step 2: Slope of the line k (m_2)

    Step 3: $m_1 \times m_2$

    Conclusion:

    b) Step 1: Rearrange the equation of the line p into the form $y = mx + c$

    Step 2: Identify the point of intersection with the y-axis

    c) Step 1: Rearrange the equation of the line k into the form $y = mx + c$

    Step 2: Identify the point of intersection with the y-axis

12. The line c has the equation $x - 3y - 6 = 0$.

    Find the equation of the line d, which is parallel to the line c and contains the point $(1, -2)$.
    Give your answer in the form $ax + by + c = 0$.

    Step 1: Find the slope of the line c

    Step 2: Find the slope of the line d, given that it is parallel to the line c

    Step 3: Equation of the line: $y - y_1 = m(x - x_1)$

    Step 4: Rearrange the equation into the required format

Exam-Style Questions

Exam Hints

1. Attempt every part of every question. You cannot get marks for a blank space!

2. Highlight key words in the question.

3. Write the relevant formula from the log tables. The correct formula with some relevant substitution will get you marks.

4. If you are asked to give your answer in a particular format and are unsure of how to do this, do as much as you can.

5. Marks are not just given for the correct answer, they are also available for work completed along the way.

Question 1 – 2022

Linked Question – Co-ordinate Geometry, Geometry, Algebra

The co-ordinate diagram below shows part of the N22 road in County Cork.

Two points on the road, P and Q, are marked on the diagram.

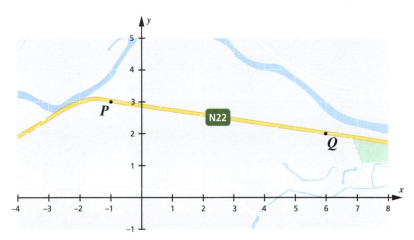

Remember!

When labelling a point, you go to the x-axis first and then the y-axis.

a) The point Q has co-ordinates (6, 2).

  Write down the co-ordinates of the point P.

  $P = (\ \ ,\ \ )$

104 SKILLS FOR EXAM SUCCESS MATHS

b) The equation of the line PQ is:

x + 7y = 20

Using this, or otherwise, find the co-ordinates of the point where the line PQ crosses the y-axis.

Answer = (,)

Remember! All the way along the y-axis, $x = 0$.

c) A new road is being built through the point Q (6, 2).

On the co-ordinate diagram, it will be a straight line segment which is **perpendicular** to PQ.

Work out the equation of this new road.

Give your answer in the form $ax + by + c = 0$, where $a, b, c \in \mathbb{R}$.

d) The distance |PQ| on the diagram is 7·1 cm, correct to 1 decimal place.

5 mm on the diagram represents 100 m.

Use this to work out the **actual** distance from P to Q. Give your answer in km.

Remember!
10 mm = 1 cm
100 cm = 1 m
1000 m = 1 km

1. Find the slope of PQ.
2. Perpendicular lines have slopes that are the 'opposite, inverse of each other'.
3. $y - y_1 = m(x - x_1)$
Remember that marks are awarded for each correct step so do as much as you can.

CHAPTER 7: CO-ORDINATE GEOMETRY OF THE LINE

Question 2 – 2022

The line h has a slope of 4 and passes through the point (20, 12).

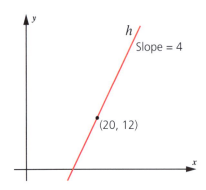

1. Use the information given to find the equation of the line.
$$y - y_1 = m(x - x_1)$$
2. The line cuts the x-axis.
3. All the way along the x-axis, $y = 0$.
4. Sub $y = 0$ into the equation of the line formula and find x.
5. You now have a point (x, y).

Find the co-ordinates of another point on the line h, other than the point (20, 12).

Show your working out.

Answer = (,)

Question 3

Linked Question – Co-ordinate Geometry, Area and Volume, Transformation Geometry, Geometry, Algebra

The co-ordinate diagram below shows the right-angled triangle ABC.

The point A has co-ordinates (4, 2).

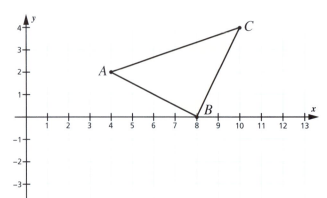

Remember!
When labelling a point, you go to the x-axis first and then the y-axis.

a) Write down the co-ordinates of the point B and the point C.

$B = (,)$ $C = (,)$

b) The table below shows the equation of the line AC and BC, where $m, k \in \mathbb{Q}$.
Work out the value of m and the value of k.

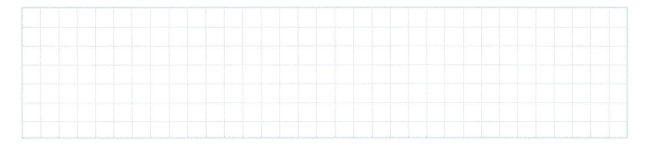

| The point C from above is (x, y). Sub the values for x and y into the equation $y = mx + \frac{2}{3}$ and solve for m. | **Line AC**
 Equation: $y = mx + \frac{2}{3}$
 Answer: $m = $ _____ | **Line AB**
 Equation: $y = \frac{-1}{2}x + k$
 Answer: $k = $ _____ | The point B above is (x, y). Sub the values for x and y into the equation $= -\frac{1}{2}x + k$ and solve for k. |

c) Show that the area of the triangle ABC is 10 square units.

Remember!
Area of a triangle $= \frac{1}{2}BH$

Base $= |AB|$

Height $= |BC|$

d) The triangle $A'B'C'$ is the image of the triangle ABC under an axial symmetry in the x-axis.

Draw the triangle $A'B'C'$ on the co-ordinate diagram.

e) Which of the points A, B or C is the intersection of the two triangles ABC and $A'B'C'$?

Intersection = where the two triangles overlap/the point the two triangles have in common.

Answer = _____

f) The point (p, s) lies inside the triangle ABC, where $p, s \in \mathbb{R}$.

Use this fact to write down the co-ordinates of a point that **must lie inside** the triangle $A'B'C'$, giving your answer in terms of p and s.

Answer = (,)

g) Use the theorem of Pythagoras to investigate if the triangle [ABC] is a right-angled triangle.

Remember!
Pythagoras' Theorem: $h^2 = a^2 + b^2$

$h = |ac|$

$a = |ab|$

$b = |bc|$

Length $= \sqrt{(x_2 - x_1)^2 + (y_2 - y_1)^2}$

CHAPTER 7: CO-ORDINATE GEOMETRY OF THE LINE 107

h) Verify, using algebra, that the point of intersection of the line AC and AB is (4,2)

1. Need the equation of the line *AC* and the equation of the line *AB* in the form $ax + by = c$.
2. Use simultaneous equations to find the point of intersection of the two lines.

Summary review:

- The formulae to find distance, midpoint, slope and equation of a line are in the *Formulae and Tables* booklet (p. 18). You *do not need to remember these.*
- You do need to remember that when the line is in the form $ax + by + c = 0$, then:
  - Slope $= -\frac{a}{b}$
  - a = sign and number in front of the x, b = sign and number in front of the y
  - If you rearrange the equation of the line into the form $y = mx + c$, then the number in front of the x is the slope (m) of the line and c is the y-intercept
- Parallel lines have the same slope.
- To find a perpendicular slope 'invert the slope and change the sign'.
- To prove two lines are perpendicular to each other:

$$m_1 \times m_2 = -1$$

(Slope of the first line × slope of the second line) = −1
- When testing if a point is on a line, if Left-hand side (LHS) = Right-hand side (RHS), the point is on the line; if LHS ≠ RHS, the point is not on the line.
- When graphing lines:
  - You only need two points to be able to draw a straight line.
  - All the way along the x-axis, $y = 0$.
  - By letting $y = 0$, you can find the point on the x-axis that is on the line (_ , 0).
  - All the way along the y-axis, $x = 0$.
  - By letting $x = 0$, we can find the point on the y-axis that is on the line (0, _).

Scan this code to access a video with worked solutions for every question.

Chapter 8: Number Patterns

The topics and skills I will work on are:

Linear or arithmetic sequences
- start term
- common difference
- general term of a linear sequence: $T_n = a + (n-1)d$

Using graphs to solve problems

Quadratic patterns/quadratic sequences
- general term of a quadratic sequence: $T_n = an^2 + bn + c$

Exponential sequences

Section A: Linear or Arithmetic Sequences

- A linear or arithmetic sequence is one in which we have a set of numbers and the difference between consecutive numbers or terms is the same (constant). For example: 2, 5, 8, 11, 14…

- In this sequence, the 1st term $(T_1) = 2$, the 2nd term $(T_2) = 5$, the 3rd term $(T_3) = 8$ and so on. The difference (d) between consecutive terms is 3.

- To generate this sequence of numbers, a formula was used.

- To create the formula, we use the following: $T_n = a + (n-1)d$, where a is the first term in the sequence and d is the difference between consecutive terms. Here, $a = 2$ and $d = 3$ so:

$T_n = a + (n-1)d$

$T_n = 2 + (n-1)3$

$T_n = 2 + 3n - 3$

$T_n = 3n - 1$. This is called the general term T_n.

- So if you want to calculate, for example, the 50th term in the sequence ($n = 50$), instead of counting up, you can use this formula to find it directly.

$T_n = 3n - 1$

$T_{50} = 3(50) - 1$

$T_{50} = 149$

Worked example

The first four terms of a sequence are 3, 7, 11, 15.

a) Prove that this is a linear sequence.

The difference between each term is 4. The 4s are referred to as the first difference. When the first difference is constant, this proves the sequence is linear.

> This statement is very important. It proves that you understand what having the first difference constant means.

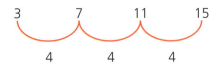

b) Identify a (the 1st term) and d (the common difference).

$a = 3 \qquad d = 4$

c) Find, in terms of n, an expression for T_n, the nth term.

$T_n = a + (n - 1)d$ — This formula is in the *Formulae and Tables* booklet (p. 22).

$T_n = 3 + (n - 1)4$ — Substitute in $a = 3$ and $d = 4$.

$T_n = 3 + 4n - 4$ — Multiply out.

$T_n = 4n - 1$ — Simplify. Now that we have the general term, we can calculate any term of the sequence.

d) Hence, or otherwise, find T_{20} the 20th term of the sequence.

$T_n = 4n - 1$

$T_{20} = 4(20) - 1 = 79$ — The 20th term of the sequence is 79.

e) Which term of the sequence is 175?

$T_n = 4n - 1$ — Here we are trying to find n, what number term of the sequence 175 is.

$4n - 1 = 175$

$4n = 175 + 1$

$4n = 176$

$n = \frac{176}{4} = 44$ — The 44th term in the sequence is 175.

110   SKILLS FOR EXAM SUCCESS MATHS

Section B: Using Graphs to Solve Problems

1. Mikolaj's parents open a savings account for him with a credit union. They lodge €20 into the account. Mikolaj decides that he will save €10 of his wages from his part-time job every week.

   a) Complete the table below.

Week number	Amount saved
0	20
1	30
2	40
4	
6	
8	
10	
12	
15	

   Remember!
   When graphing functions:
   1. Always use a pencil.
   2. Label the x- and the y-axes.
   3. Make sure the scales along the x- and y-axes are constant.
   4. Plot the x co-ordinate first and then the y. Plot the points clearly and accurately.
   5. Join your points together. There are marks for plotting and for joining the points.

   b) Represent this information on a graph.

   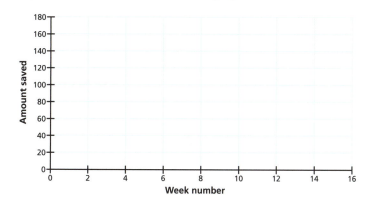

   c) How much money will Mikolaj have in his account after 13 weeks?

   To calculate the amount of money in the account after 13 weeks, you go to 13 on the x-axis, draw a vertical line up until you hit the graph and then draw a horizontal line over until you hit the y-axis.

   d) Looking at the data, write an expression in terms of n for T_n, the amount of money in the account after n weeks.

   $a = 20 \quad d = 10$

   Exam Hint
   Always write the formula that you are using. The correct formula with any correct relevant substitution will merit marks in the exam.

   e) i. How much money will be in the account after 25 weeks?

   $T_n = 10n + 20$

CHAPTER 8: NUMBER PATTERNS

ii. How much money will Mikolaj have saved after 25 weeks?

f) Mikolaj is saving for a new games console, which costs €600.

How many weeks will it take him to save this money?

$T_n = 10n + 20$

g) Calculate the slope of the line.

Give your answer using $y = mx + c$ and explain what it means in the context of this question.

Exam Hint

Always write the formula that you are using. The correct formula with any correct relevant substitution will merit marks in the exam.

Remember!

Slope $= \dfrac{y_2 - y_1}{x_2 - x_1}$ or $\dfrac{\text{rise}}{\text{run}}$

h) Investigate if (20, 220) is on the line.

i) What does the point (20, 220) mean in the context of this question?

Section C: Quadratic Patterns/Quadratic Sequences

- A quadratic sequence is one in which we have a set of numbers and the second difference is constant.

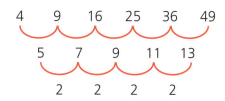

  The first difference is not constant, so it is not a linear sequence.

  The second difference is constant.

  This means that it is a quadratic sequence.

- Quadratic sequences also include n^2.

- To find the general term of a quadratic sequence we use the following formula:

  $T_n = an^2 + bn + c$

- Graph of a quadratic sequence.

Exam Hint

This formula is not in the *Formulae and Tables* booklet and needs to be remembered.

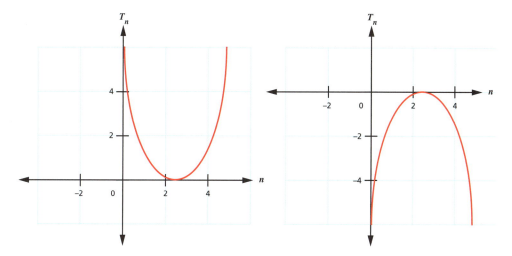

Remember!

The formula for a quadratic function is:

$f(x) = ax^2 + bx + c$

Note the similarity to the T_n formula.

CHAPTER 8: NUMBER PATTERNS

Worked example

a) Find the general term of the sequence 3, 7, 15, 27, 43.

Step 1: We need to identify what type of a sequence this is.

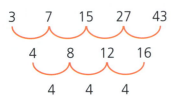

The second difference is constant. This means that it is a quadratic sequence.

Exam Hint
Remember there are marks available for each step, not just the final answer. So even if you are unsure how to calculate the general term, you can still identify what type of a sequence you are dealing with.

Step 2: Find the general term of a quadratic sequence.

$T_n = an^2 + bn + c$

To find the value of a (the number in front of n^2), we halve the second difference.

a is half the value of the common difference.

Here the second difference is 4 so $a = 2$.

$T_n = 2n^2 + bn + c$

Step 3: We now have to find the value of b and c.

Because there are two unknowns here, b and c, we need to form two simultaneous equations to find the two unknown values.

$T_1 = 3$ and $T_2 = 7$ are used below. You can use any of the terms T_3, T_4, T_5, etc. However, I recommend using the smaller values as this reduces the risk of mathematical errors.

The 1st term in the sequence, $T_1 = 3$

$2(1)^2 + b(1) + c = 3$

$2 + b + c = 3$

$b + c = 3 - 2$

$b + c = 1$

The 2nd term of the sequence, $T_2 = 7$

$2(2)^2 + b(2) + c = 7$

$8 + 2b + c = 7$

$2b + c = 7 - 8$

$2b + c = -1$

Step 4: We will now use simultaneous equations to solve for b and c.

$b + c = 1$ $\times -1$ $-b - c = -1$ $-b - c = -1$

$2b + c = -1$ $2b + c = -1$ $2b + c = -1$

                                                       $b = -2$

We now know that $b = -2$. We must now find c.

$b + c = 1$

$(-2) + c = 1$

$c = 1 + 2$

$c = 3$

General term, $T_n = 2n^2 + bn + c$

$2n^2 - 2n + 3$

b) Hence, or otherwise, find the 20th term of the sequence.

$T_n = 2n^2 - 2n + 3$

$T_{20} = 2(20)^2 - 2(20) + 3$

$= 763$

2. A small rocket is fired into the air from a fixed position on the ground.

   The time and its height above the ground are recorded in the table below.

Time (secs)	0	1	2	3	4	5	6
Height (m)	0	9	16	21	24	25	24

   a) Draw a graph to represent the height of the rocket after 6 seconds.

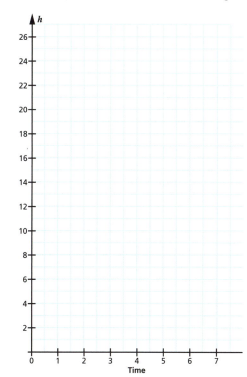

Remember!

When graphing functions:
1. Always use a pencil.
2. Label the x- and the y-axes.
3. Make sure the scales along the x- and y-axes are constant.
4. Plot the x co-ordinate first and then the y. Plot the points clearly and accurately.
5. Join your points together. There are marks for plotting and for joining the points.

CHAPTER 8: NUMBER PATTERNS   115

b) Find the height of the rocket after 4·5 seconds.

c) At what time is the rocket 18 m above the ground?

d) Is this an example of a linear or a quadratic sequence? Justify your answer.

Justify means write a sentence or show how you got your answer or reached your conclusion.

e) Find T_n, the general term of this sequence.

Hint: $T_n = an^2 + bn + c$ where a = half of the second difference.

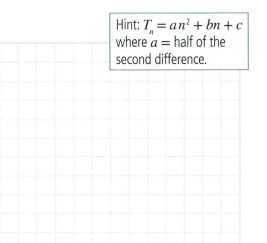

Section D: Exponential Sequences

- An exponential sequence is one in which we have a set of numbers that is generated by multiplication of a constant term. For example: 1, 3, 9, 27, 81, …

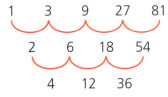

The first difference is not constant, so it is not a linear sequence.

The second difference is not constant, so it is not a quadratic sequence.

We now check if it is exponential. Each term is generated by multiplying the previous one by 3, therefore it is exponential.

116 SKILLS FOR EXAM SUCCESS MATHS

3. The number of bacteria growing on a Petri dish can be represented by the following:

   $T_n = 2^n$, where n represents the number of stages the bacteria are allowed to reproduce for.

   a) Complete the table below.

Stage	0	1	2	3	4	5	6
Number of bacteria							

   b) Draw a graph to represent this information.

   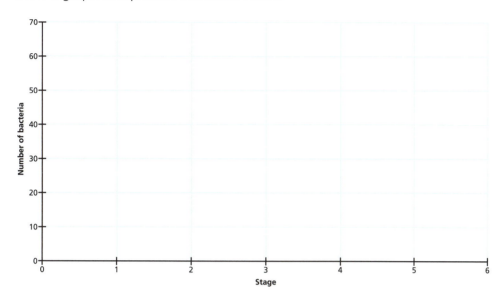

   c) Is this an example of a linear, quadratic or exponential sequence?

   Justify your answer.

   Justify means write a sentence or show how you got your answer or reached your conclusion.

   d) If the pattern was to continue, how many bacteria would be present after 20 stages?

   Give your answer in the form $a \times 10^n$, where $n \in \mathbb{Z}$ and $1 \leq a < 10$, correct to 3 significant figures.

CHAPTER 8: NUMBER PATTERNS

Exam-Style Questions

Exam Hints

1. Attempt every part of every question. You cannot get marks for a blank space!
2. Highlight key words in the question. Make sure you understand what you are being asked to do.
3. Write the relevant formula. The correct formula with some relevant substitution will get you marks.
4. If you are asked to give your answer in a particular format and are unsure of how to do this, do as much as you can.
5. Marks are not just given for the correct answer, they are also available for work completed along the way.

Question 1

Linked Question – Number Patterns, Graphing Functions, Arithmetic, Distance/Time/Speed, Algebra, Co-ordinate Geometry of the Line

A taxi fare is calculated based on the distance travelled.

Company A charges an initial fee of €5 for journeys up to one kilometre and €1 for every kilometre after that.

Company B does not have a standing charge but charges €2 per kilometre.

a) Complete the table below:

Distance	1 km	2 km	3 km	4 km	5 km	6 km	8 km	10 km
Company A	5		7				12	
Company B	2	4				12		

b) Represent this information on the graph below.

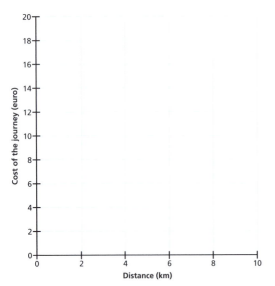

Remember!

When graphing functions:

1. Always use a pencil.
2. Label the x- and the y-axes.
3. Make sure the scales along the x- and y-axes are constant.
4. Plot the x co-ordinate first and then the y. Plot the points clearly and accurately.
5. Join your points together. There are marks for plotting and for joining the points.

SKILLS FOR EXAM SUCCESS MATHS

c) Are these both linear sequences? Justify your answer.

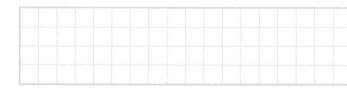

Exam Hint
This is a two-part question. You are asked to give your answer **and** to justify your answer. There are marks available for both parts, so don't forget to answer both!

d) If you had to travel a distance of 5 km, which company would you choose to travel with? Justify your answer.

Which company?

_____

Justification:

e) Zalia needs to travel to the airport, a distance of 60 km from her home.

How much money will she save travelling with Company A?

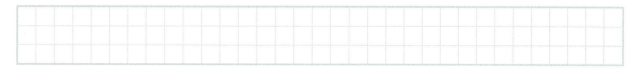

f) The driver drives at a speed of 80 km per hour to the airport.

How long, in minutes, does the journey to the airport take?

g) The car consumes 5·6 litres of diesel for every 100 km travelled. Diesel costs €1·92 per litre. Find the cost of fuel to the driver.

CHAPTER 8: NUMBER PATTERNS   119

h) Verify, using algebra, that the point of intersection of the lines in part (b) is (4, 8).

1. Find the slope of both lines.
2. Find the equation of both lines.
3. Use simultaneous equations to find the point of intersection of the two lines.

Question 2

Linked Question – Number Patterns, Geometry, Applied Measure

The following sequence of patterns is created using matchsticks to form equilateral triangles.

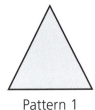
Pattern 1

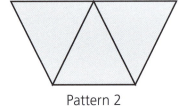
Pattern 2

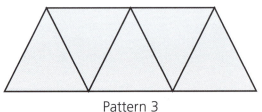
Pattern 3

a) Draw the next pattern of the sequence.

Exam Hint
Always use a pencil to draw diagrams.

b) Complete the table below to find the number of matchsticks required to make each of the first six patterns of the sequence.

Pattern number	1	2	3	4	5	6
Number of matchsticks	3	7				

c) Is this a linear, quadratic or exponential sequence?

Justify your answer.

Justify means write a sentence or show how you got your answer or reached your conclusion.

120   SKILLS FOR EXAM SUCCESS MATHS

d) Find T_n, the general term of the sequence.

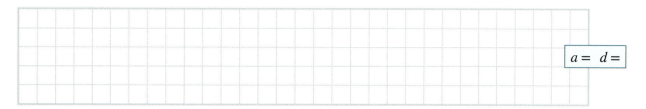

$a = \quad d =$

e) Hence, or otherwise, find T_{18}, the 18th term in the sequence.

f) The area of each triangle is $5\sqrt{3}$ cm². Find, correct to the nearest cm², the combined area covered by the first five patterns in the sequence.

g) Construct the equilateral triangle ABC, each side being 6 cm.

Exam Hint
Always use a pencil to draw diagrams.

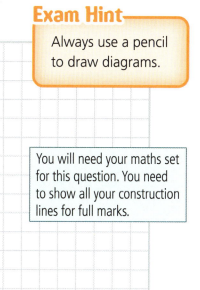

You will need your maths set for this question. You need to show all your construction lines for full marks.

h) Construct the perpendicular bisector of the line segment [AB].

Extend the perpendicular bisector to the vertex C.

Where the perpendicular bisector cuts [AB], label this the point D.

CHAPTER 8: NUMBER PATTERNS

i) Prove that the triangle *ADC* is congruent to the triangle *DBC*.

Requirements for congruency:
SSS or
SAS or
ASA or
RHS.

Question 3

Linked Question – Number Patterns, Graphing Functions

a) The first five numbers in a pattern are given in the table below.

Term	Number
T_1	13
T_2	15
T_3	19
T_4	25
T_5	33
T_6	
T_7	
T_8	

i. Follow the pattern to write the next three terms into the table.

ii. Use the data in the table to show that the pattern is quadratic.

b) Represent the data on the graph below.

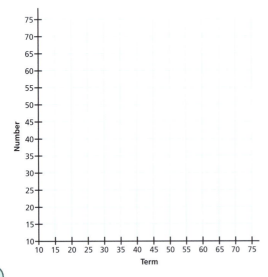

122 SKILLS FOR EXAM SUCCESS MATHS

c) $T_n = an^2 + bn + c$ where b and $c \in \mathbb{Z}$. Find the value of b and the value of c.

Remember!
$\mathbb{Z}$ = integers which are all the positive and negative whole numbers.

d) Hence, or otherwise, find T_{10}, the 10th term in the sequence.

Do not rule out attempting this part of the question if you did not get the formula in part (c). You can count up to the 10th term, even if you are unable to complete part (c).

Summary review:

- $T_n = a + (n - 1)d$ is the general term for a linear sequence. This formula is in the *Formulae and Tables* booklet (p. 22) so you don't need to remember it.

- $T_n = an^2 + bn + c$ is the general term for a quadratic equation. This formula is **not** in the *Formulae and Tables* booklet.

    - The value of a (the coefficient of n^2) is calculated by finding half of the second difference.

    - b and c are evaluated using simultaneous equations.

- You need to remember the conditions necessary to be a linear sequence (first difference constant), quadratic sequence (second difference constant) and an exponential sequence (generated by multiplication).

Scan this code to access a video with worked solutions for every question.

CHAPTER 8: NUMBER PATTERNS

Chapter 9: Functions and Graphing Functions

The topics and skills I will work on are:

Functions ☐
- Identifying functions ☐
- Representing functions with a mapping diagram and as a set of couples ☐
- Understanding domain and range ☐

Graphing linear functions ☐

Graphing quadratic functions ☐
- Quadratic sequences ☐
- Quadratic functions ☐

Drawing quadratic functions ☐

Graphing exponential functions ☐
- Interpreting graphs ☐
- Using graphs to solve real-life problems ☐
- Functions with missing coefficients ☐
- Creating and interpreting distance/time/speed graphs. ☐

Section A: Functions

- A function is an expression, rule or law that defines a relationship between one variable (the independent variable) and another variable (the dependent variable).

- A function produces only one output number for each input value.

- In order to be classified as a function, the following rules apply:
  1. Every input maps to only one output.
  2. Every input must be used.

Remember!

You will need to remember the following terms:
- Domain: the set of numbers that are put into the function
- Co-domain: all the possible outputs
- Range: the set of numbers that come out of the function.

1. Identify if each of the following are functions or not. In each case give a reason for your answer.

a) Input/Domain → Output/Range

b) Input/Domain → Output/Range

c) Input/Domain → Output/Range

Answer: _____   Answer: _____   Answer: _____

Reason: Reason: Reason:

2. The function f is defined as $f : x \to x^2 - 2x - 15$, where $x \in \mathbb{R}$. Find:

a) i. $f(-2)$ ii. $f(-1)$ iii. $f(0)$
   iv. $f(1)$ v. $f(2)$

Worked example

i. $f(-2)$

$f(-2)$ $x^2 - 2x - 15$

$= (-2)^2 - 2(-2) - 15$

$= -7$

To find $f(-2)$, you must sub -2 into the equation for each x.

Remember!
A function can be recorded in a variety of different ways. For example:
$f(x) =$
$F : x \to$
$y =$
$T_n =$

ii. $f(-1)$

$(\ )^2 - 2(\ ) - 15$

Remember!
Always be aware of your signs:
$(+)(+) = +$ $(+)(-) = -$
$(-)(-) = +$ $(-)^2 = +$

iii. $f(0)$

CHAPTER 9: FUNCTIONS AND GRAPHING FUNCTIONS

iv. $f(1)$

v. $f(2)$

b) Represent f with a mapping diagram, clearly labelling the domain and the range.

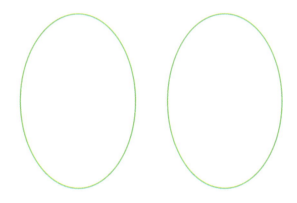

c) Find the values for x for which $f(x) = 0$.

$x^2 - 2x - 15 = 0$, solve for x.
$(x \quad)(x \quad) = 0$

Hint: Factorise the equation. You need two numbers that multiply to give -15 and the same two numbers must add together to give -2.

Section B: Graphing Linear Functions

- In these questions, the domain is your x values.
- A domain of $-2 \leq x \leq 4$ means your values for x are $-2, -1, 0, 1, 2, 3, 4$ and you must find their corresponding y values. We discussed this earlier when we covered linear sequences.

$+x$ means the graph will be increasing.
$-x$ means the graph will be decreasing.

3. a) On the same axes and scales, graph the functions $f:x \to 2x - 3$ and $g:x \to 1 - 2x$ in the domain $-2 \leq x \leq 4$, where $x \in \mathbb{R}$.

Step 1: We have to identify two points on each line in order to be able to graph it. In the domain $-2 \leq x \leq 4$, where $x \in \mathbb{R}$, this means that the lines must be drawn between $x = -2$ and $x = 4$.
We will use this information to find the two points on each line.

x	$f(x) = 2x - 3$	y	Point
-2	$2(-2) - 3$	-7	$(-2, -7)$
4	$2(4) - 3$	5	$(4, 5)$

x	$g(x) = 1 - 2x$	y	Point

Step 2: Graph both lines.

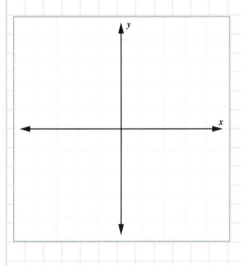

Remember!

When graphing functions:
1. Always use a pencil.
2. Label the x- and the y-axes.
3. Make sure the scale along the x-axis is constant.
4. Make sure the scale along the y-axis is constant.
5. Remember to plot the x co-ordinate first and then the y. For example, for the point $(-2, -7)$, go to -2 on the x-axis first, then -7 on the y-axis.
6. Plot the points clearly and accurately.
7. Finally, join your points together. There are marks available for plotting the points correctly **and** for joining them together.

b) On your graph, identify the point of intersection of the two lines and write in the co-ordinates of this point.

Step 3: Identify the point of intersection on the graph.
Using a pencil and a neat arrow, identify the point of intersection and label the point on the graph.
Point of intersection from the graph: _____

When two graphs intersect, they have the same values for x and y. Because we have two unknown values (x **and** y), we use simultaneous equations to find the point of intersection.

CHAPTER 9: FUNCTIONS AND GRAPHING FUNCTIONS

c) Use algebra to verify the point of intersection of the two lines whose equations are $y = 2x - 3$ and $y = 1 - 2x$.

> Step 4: Use algebra to verify the point of intersection.
>
> Here you will use simultaneous equations to confirm the point of intersection.
> In this situation you need to rearrange the equations so that they are in the form $ax + by = c$.
>
> $y = 2x - 3$ $\qquad\qquad y = 1 - 2x$
>
> $2x - y = 3$ Equation 1 $\qquad\qquad$ _____ Equation 2
>
> Once in the required format, you can solve for x and y using simultaneous equations.

Remember!
Steps for solving simultaneous equations:
1. Use the elimination method to cancel one of the variables.
2. Find the value of one variable.
3. Use substitution to find the value of the other variable.

Section C: Graphing Quadratic Functions

- A quadratic function takes the form $f(x) = ax^2 + bx + c$.

- To find the roots of the function, we factorise the equation to solve for x. This tells us where the curve cuts through the x-axis.

- The value of c tells us where the curve cuts the y-axis.

Quadratic Sequences

- To find the general term of a quadratic sequence we use $T_n = an^2 + bn + c$. This is not in the *Formulae and Tables* booklet so you need to remember it.

Remember!
A quadratic pattern/sequence is one in which the second difference is constant between the numbers in the sequence.

Graph of a quadratic sequence

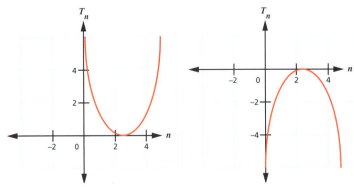

128 SKILLS FOR EXAM SUCCESS MATHS

Quadratic Functions

Often in the exam, you will be given a graph or asked to draw a graph and then asked to answer some questions from the graph. This section covers some of the main types of questions you will be asked.

Identify whether the graph is positive or negative

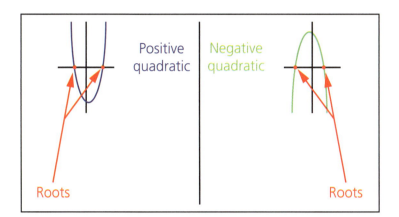

Quadratic functions can take two forms:
If the coefficient of x^2 is positive (i.e. $+ax^2$), a graph like the first one here (a u-shape/smiley face) will be produced.
If the coefficient of x^2 is negative (i.e. $-ax^2$), a graph like the second one here (an n-shape/sad face) will be produced.

Find the roots of the equation/find the values for x for which $f(x) = 0$

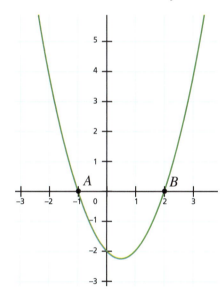

Here you are being asked 'where does the curve cut the x-axis?'
From the example, you can see that the curve cuts the x-axis in two places, at $x = -1$ and $x = 2$.
Alternatively, you may be given the equation of the curve and be asked to find the roots/solve for x.
Equation of the curve: $x^2 - x - 2 = 0$

$(x + 1)(x - 2)$ Factorise the equation.
$x + 1 = 0$ $x - 2 = 0$ Put each factor equal to 0.
$x = -1$ $x = 2$ Solve for x.

Find the values for x for which $f(x) = 3$

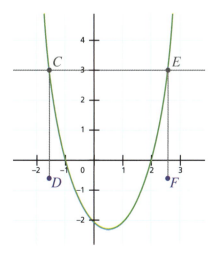

$f(x) =$ is another way of writing $y =$.
You are being asked if $y = 3$, what are the corresponding values for x?
With your ruler, draw the line $y = 3$.
Where the line cuts the curve, you come down and read off the corresponding values for x.
In this example, when $f(x) = 3$, $x = -1.8$ and $x = 2.8$.

CHAPTER 9: FUNCTIONS AND GRAPHING FUNCTIONS

Find the value of $f(2·4)$

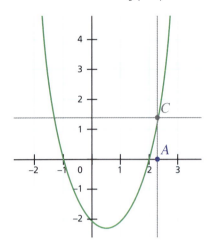

$f(2·4)$ is asking you when $x = 2·4$, what is the corresponding value for y?
Go to $2·4$ on the x-axis.
Use your ruler to draw a straight line to the curve.
Draw a straight line to the y-axis and read off the corresponding value for y.
In this example, when $x = 2·4$, $y = 1·4$.

Maximum and minimum points (x, y), maximum and minimum values

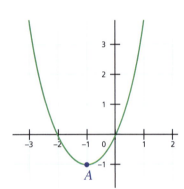

In this graph, the min point is $(-1, -1)$.
The minimum value of the function is the lowest value on the y-axis, -1.

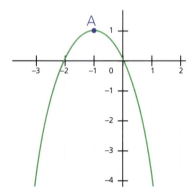

In this graph, the max point is $(-1, 1)$.
The maximum point is the highest value on the y-axis, 1.

Find the values for x when $f(x)$ is increasing and $f(x)$ is decreasing

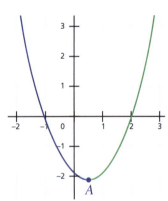

In this graph, the curve is decreasing between $x = -2$ and $x = 0·5$.
We write this as $-2 \leq x \leq 0·5$.
Here the curve is increasing between $x = 0·5$ and $x = 3$.
We write this as $0·5 < x \leq 3$.

Find the values for x when $f(x) > 0$ (positive) and $f(x) < 0$ (negative)

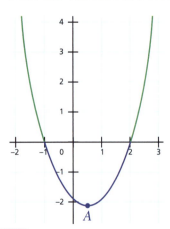

$f(x) > 0$ is looking for where the graph is positive, where it is above the x-axis.
In this example, the graph is above the x-axis in two places, between $x = -2$ and $x = -1$ and between $x = 2$ and $x = 3$.
We write that as follows: $-2 \leq x < -1$ and $3 < x \leq 4$.
$f(x) < 0$ is looking for where the graph is negative, where it is below the x-axis.
In this example, the graph is below the x-axis between $x = -1$ and $x = 2$ or $-1 < x < 2$.

SKILLS FOR EXAM SUCCESS MATHS

Axis of symmetry

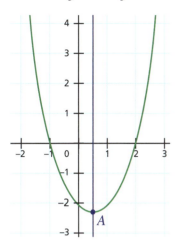

The axis of symmetry is the line that divides the curve exactly in half.
In this example, the equation of the axis of symmetry is $x = 0.5$.

Two functions graphed on the same axes and scales

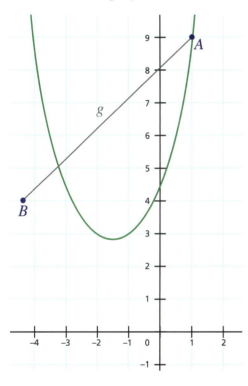

The green curve is $f(x)$ and the black line is $g(x)$.
Where is $f(x) = g(x)$? This is asking where the line and the curve intersect: $(-3, 5)$ and $(1, 9)$.
Where is $f(x) > g(x)$? This is asking for the values of x where the curve is above the line. The curve is above the line in two places: between $x = -5$ and $x = -3$ and $x = 1$. We write this as $-5 \leq x < -3$ and $x \leq 1$.
Where is $f(x) < g(x)$? This is asking for the values of x where the curve is below the line. The curve is below the line between $x = -3$ and $x = 1$. We write this as $-3 \leq x < 1$.

Section D: Drawing Quadratic Functions

- The domain are your x values. In the domain $-2 \leq x \leq 3$, your values for x are $-2, -1, 0, 1, 2, 3$ and you must find their corresponding y values.

- $g: x \rightarrow x - 1$ is a linear equation (x). It will generate a straight line.

- $f: x \rightarrow 2x^2 - x - 6$ is a quadratic function (x^2). It will generate a curve.

- You need to be familiar with the settings on your own calculator to input data correctly.

CHAPTER 9: FUNCTIONS AND GRAPHING FUNCTIONS

4. a) Using the same axes and scales, draw the graphs of the functions $g:x \to x - 1$ and $f:x \to 2x^2 - x - 6$, in the domain $-2 \leq x \leq 3$.

Step 1: Find the points necessary to draw $f:x$ and $g:x$.

$g:x \to x - 1$ —— As we are dealing with a straight line here, we only need two points to draw the line.

x	$x - 1$	y	Point
-2			
3			

$f:x \to 2x^2 - x - 6$ —— We can already tell from the equation that the curve will cut the y-axis at $(0, -6)$.

Exam Hint

Although it is possible to do all necessary calculations on your calculator, I would highly recommend including a table like this so that the examiner can clearly see where you got each point. Also, if you make a mistake, the examiner will be able to pinpoint where the mistake was made, which means you will lose the minimum amount of marks for your error.

As this is a quadratic equation, we need to find all the points in the domain to plot it correctly.

x	$2x^2 - x - 6$	y	Point
-2			
-1			
0			
1			
2			
3			

Step 2: Draw the graph.

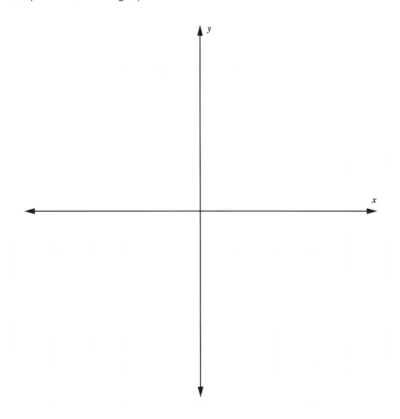

b) Answer the questions that follow based on the above graphs:

  i. Find the values for x for which $f(x) = 0$ _____

  ii. Find the values for x for which $f(x) = 2$ _____

  iii. Find the value of $f(-0.5)$ _____

  iv. What is the minimum value of $f(x)$? _____

  v. What is the minimum point of $f(x)$? _____

  vi. Find the values of x for which:

  - $f(x)$ is increasing _____

  - $f(x)$ is decreasing _____

  vii. Where is $f(x) = g(x)$? _____

  viii. For what values of x is $f(x) < g(x)$? _____

5. a) Three functions, $f(x)$, $g(x)$ and $h(x)$, are defined as follows:

$f(x) = 2x^2 + x - 6 \qquad g(x) = x^2 - 6x + 9 \qquad h(x) = x^2 - 2x$

i.

Solve $f(x) = 0$	Solve $g(x) = 0$	Solve $h(x) = 0$

ii.

Identify where $f(x)$ cuts the y-axis.	Identify where $g(x)$ cuts the y-axis.	Identify where $h(x)$ cuts the y-axis.

b) The table below shows the sketches of six different functions. Three of the sketches belong to the three functions from part (a).

Write f(x), g(x) or h(x) into the box underneath the correct sketch for each of the three functions.

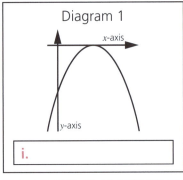

Diagram 1

i.

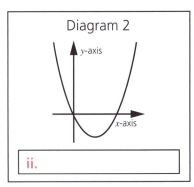

Diagram 2

ii.

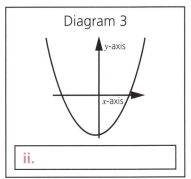

Diagram 3

iii.

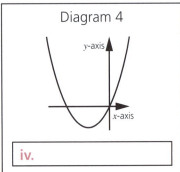

Diagram 4

iv.

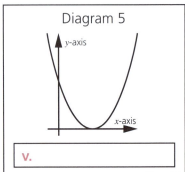

Diagram 5

v.

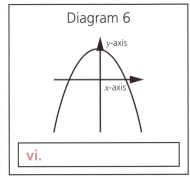

Diagram 6

vi.

Section E: Graphing Exponential Functions

- An exponential function is a function in the form $f(x) = ab^x$ for positive real numbers a and b.

- It is a mathematical function used to calculate the exponential growth or decay of a given set of data.

6. The rate of bacterial growth can be represented by the function $h: x \rightarrow 3(2^x)$, where x represents the time the bacteria are growing for.

   Draw a graph of this function in the domain $0 \leq x \leq 3$.

   Remember!
   The domain are your x values. In the domain $0 \leq x \leq 5$, your values for x are 0, 1, 2, 3, 4, 5 and you must find their corresponding y values.

   Table of values

x	$3(2^x)$	y	Point
0			
1			
2			
3			

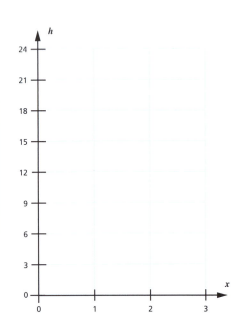

SKILLS FOR EXAM SUCCESS MATHS

7. Two mobile phone companies, Cellulon and Mobil, offer price plans for mobile internet access. A formula, in x, for the total cost per month for each company is shown in the table below. x is the number of MB of data downloaded per month.

Phone company	Total cost per month (cents)
Cellulon	$C(x) = 4x$
Mobil	$M(x) = 1000 + 2x$

a) Draw the graphs of $C(x)$ and $M(x)$ on the co-ordinate grid below to show the total cost per month for each phone company for $0 \leq x \leq 700$. Label each graph clearly.

Calculations:

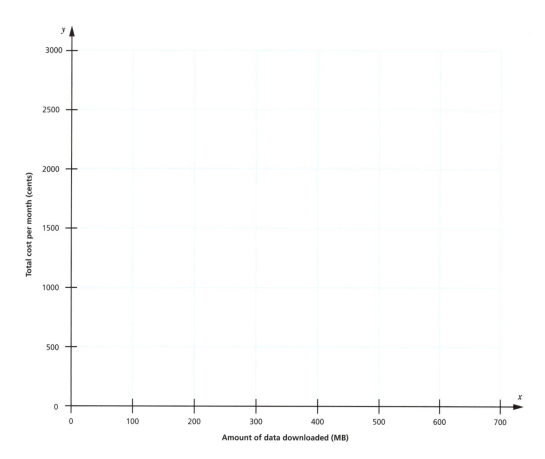

CHAPTER 9: FUNCTIONS AND GRAPHING FUNCTIONS

b) Which company charges no fixed monthly fee?

Justify your answer, with reference to the relevant formula or graph.

Answer: _____

Reason:

c) Write down the point of intersection of the two graphs.

d) Mollie wants to buy a mobile phone from one of these companies, and wants her mobile internet bill to be as low as possible.

Explain how your answer to part (c) would help Mollie choose between Cellulon and Mobil.

8. The graph below shows the distances travelled along a track by Barry over the course of a race. The graph is in two sections, labelled A and B.

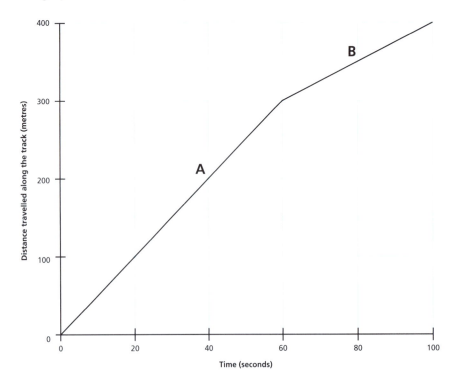

a) Show that Barry's speed in section A is 5 metres per second.

Remember!
Speed = $\frac{\text{Distance}}{\text{Time}}$

b) Find Barry's speed in section B, in metres per second.

c) Table 1 shows graphs of the distance travelled along the track by Bill, Claire and Dee during the same race. Each person's name is given next to their graph.

Table 2 shows graphs of the speed of Bill, Claire and Dee during the race.

Complete Table 2, by writing the correct name next to each graph.

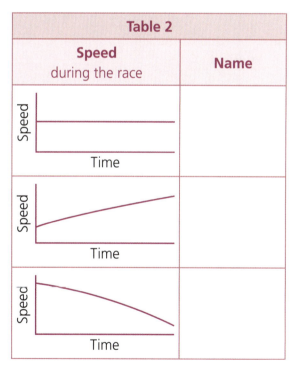

CHAPTER 9: FUNCTIONS AND GRAPHING FUNCTIONS   137

Exam-Style Questions

Exam Hints

1. Attempt every part of every question. You cannot get marks for a blank space!
2. Highlight key words in the question. Make sure you understand what you are being asked to do.
3. Write the relevant formula. The correct formula with some relevant substitution will get you marks.
4. If you are asked to give your answer in a particular format and are unsure of how to do this, do as much as you can.
5. Marks are not just given for the correct answer, they are also available for work completed along the way.

Question 1

Linked Question – Graphing Functions, Co-ordinate Geometry

A school can get its electricity from one of two companies, Buzz or Lecky. The graphs below show the cost of the electricity per month from each company if the school uses x units of electricity. The cost from Buzz is $b(x)$ and the cost of Lecky is $l(x)$.

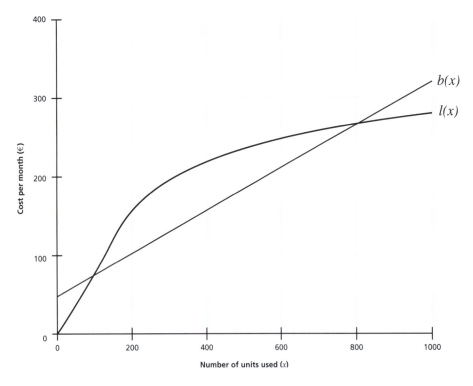

One of the companies charges a fixed fee each month, plus a fee for each unit of electricity used.

a) State which company charges no fixed fee.

Give a reason for your answer, based on the graph.

Exam Hint

This is a two-part question. You are asked to give your answer **and** give a reason for your answer. There are marks available for both parts so don't forget to answer both!

138 SKILLS FOR EXAM SUCCESS MATHS

b) Write the domain and the range of the function b(x), as shown in the graph.

Domain = _____
Range = _____

> **Remember!**
> Domain: The set of numbers that are put into the function.
> Range: The set of numbers that come out of the function

c) **i.** From the graph, estimate the points of intersection of b(x) and l(x).

> The point of intersection is where both graphs overlap.

ii. Use the graphs to estimate the set of values of $x \in \mathbb{R}$ for which $b(x) < l(x)$.

> Here you are being asked for what values of x is b(x) below l(x).

iii. Explain what your answer to (c)(i) means about the cost of electricity from Buzz and Lecky.

d) **i.** Find the slope of the graph of b(x).

> Slope = $\frac{\text{Rise}}{\text{Run}}$
> Slope = $\frac{y_2 - y_1}{x_2 - x_1}$

ii. Explain what your answer to (d)(i) means about the cost of electricity from Buzz.

e) **i.** Where does b(x) cut the y-axis?

ii. Hence, or otherwise, find the equation of b(x). Give your answer in the form $y = mx + c$.

CHAPTER 9: FUNCTIONS AND GRAPHING FUNCTIONS

Question 2

Linked Question – Graphing Functions, Algebra

The function $h(x)$ below gives the approximate height of the water at Howth Harbour on a particular day, from 12 noon to 5 pm

$h(x) = 10x^2 - 50x + 130$

where $h(x)$ is the height of the water in centimetres and x is the time in hours after 12 noon.

Exam Hint

I recommend that you write out each step clearly so that if you do make a mistake, the examiner will be able to pinpoint exactly where your error is and deduct the minimum amount of marks.

a) Draw the graph of the function

$h(x) = 10x^2 - 50x + 130$

on the axes below, for $0 \leq x \leq 5$, where $x \in \mathbb{R}$.

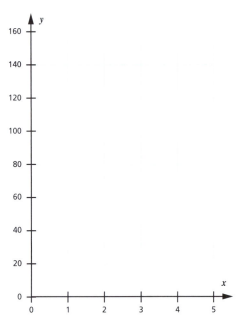

x	$10x^2 - 50x + 130$	y	Point
0			
1			
2			
3			
4			
5			

b) Use your graph from part (a) to answer the following questions.

  i. Find the height of the water at 12 noon.

  ii. Estimate the height of the water at its lowest point.

  > Estimate means to give the answer as accurately as you can.

iii. After 12 noon, how long did it take before the water was at its lowest point?

The graph on the right shows the approximate height of the water in centimetres in Crookhaven on a different day, from 12 noon to 6 pm. The graph is symmetrical.

On this day, the height of the water at 12 noon was 180 cm, and the height of the water at the lowest point on the graph was 0 cm.

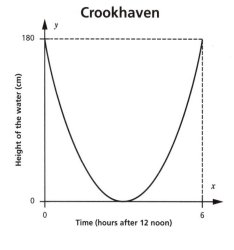

c) Taking x as the time in hours after 12 noon, this graph is given by the function

$$g(x) = ax^2 + bx + c$$

where $a, b, c \in \mathbb{Z}$ and $x \in \mathbb{R}$.

i. Find the value of c.

Hint: c is the point where the curve cuts the y-axis.

ii. Hence, or otherwise, find the values of a and b.

Remember!

Steps for solving simultaneous equations:

1. Use the elimination method to cancel one of the variables.
2. Find the value of one variable.
3. Use substitution to find the value of the other variable.

CHAPTER 9: FUNCTIONS AND GRAPHING FUNCTIONS   141

Question 3

Linked Question – Distance/Speed/Time, Number Patterns, Real Numbers

Poppy and Ella ran a 5 km race. The graph below shows the time that it took Ella to run d km during the race. One of the points on the graph is marked A.

Distance is on the horizontal axis so, for example, it took Ella 26 minutes to run the whole 5 km.

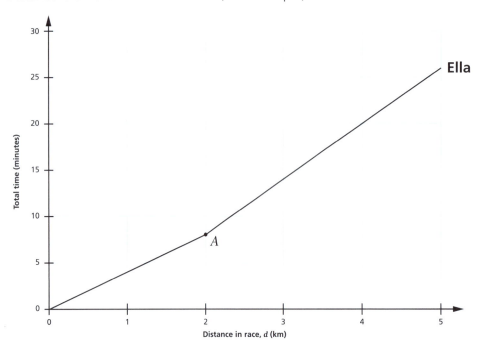

The table below shows the total time it took Poppy and Ella to run each of the given distances in the race.

a) Using the figures in the table, draw a graph on the diagram above to show the time it took Poppy to run d km during the race, $0 \leq d \leq 5$ and $d \in \mathbb{R}$.

b) Using the graph, fill in the three missing values in the table below.

Distance in the race (km)	Total time for Poppy (mins)	Total time for Ella (mins)
1	5	4
2	10	
3	17	
4	24	
5	30	26

c) Show that Poppy's times in the table do not make a quadratic sequence.

> A quadratic sequence is one where the second difference is constant.

142 SKILLS FOR EXAM SUCCESS MATHS

d) It took Ella 26 minutes to run 5 km. Work out Ella's average speed for the race.

Give your answer in kilometres per hour, correct to 2 decimal places.

> **Speed** = $\frac{\text{Distance}}{\text{Time}}$
>
> You are asked to give your answer to 2 decimal places. Don't forget to round your answer in order to get full marks.

e) Tick the correct box to show what happened to Ella's speed after 2 km, which is marked A on the graph.

Justify your answer.

Ella's speed increased ☐   Ella's speed decreased ☐   Ella's speed stayed the same ☐

Justification:

Exam Hint
There are marks available for both ticking the correct box and for the justification. So even if you find it hard to explain your answer, still tick the box you think is correct.

f) Ciarán also ran the 5 km race. He drew the graph below to show the time it took him to run d km during the race. The part of Ciarán's graph marked C is a vertical line.

What does this part tell us about Ciarán's running at this stage of the race?

Give as much detail as possible.

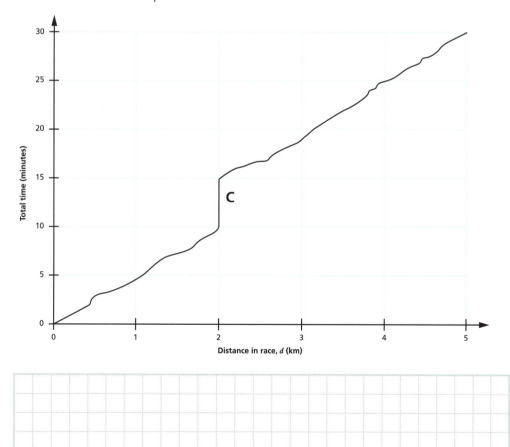

CHAPTER 9: FUNCTIONS AND GRAPHING FUNCTIONS

Chapter 10: Trigonometry

The topics and skills I will work on are:

Properties of right-angled triangles ☐
The theorem of Pythagoras ☐
Using trigonometric ratios to find missing sides in a right-angled triangle ☐
Using trigonometric ratios to find missing angles ☐
Using trigonometry to solve real-life problems. ☐

Section A: Properties of Right-Angled Triangles

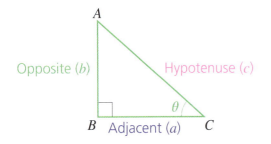

In a right-angled triangle, the sides are labelled the hypotenuse, the adjacent side and the opposite side.
The hypotenuse is the side across from the right angle.
The opposite side is across from the reference angle.
The adjacent side is adjacent to (beside) the reference angle.
It is important that these sides are labelled correctly.

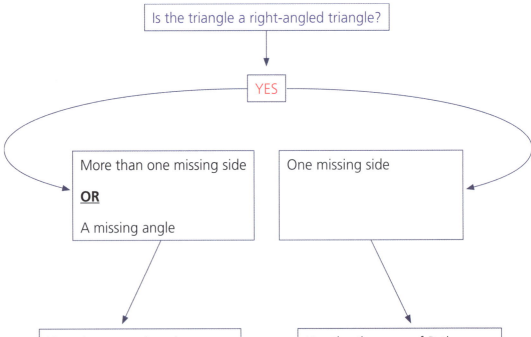

There are various memory aids to help you remember these trigonometric ratios. It is important to note that it is not enough to write the memory aids in the exam, you must present the ratios in the format that they are shown here.

Use trigonometric ratios:

$$\sin\theta = \frac{\text{opposite}}{\text{hypotenuse}}$$

$$\cos\theta = \frac{\text{adjacent}}{\text{hypotenuse}}$$

$$\tan\theta = \frac{\text{opposite}}{\text{adjacent}}$$

Use the theorem of Pythagoras

$c^2 = a^2 + b^2$. This formula is found in the *Formulae and Tables* booklet (p. 16).

144 SKILLS FOR EXAM SUCCESS MATHS

Section B: The Theorem of Pythagoras

- The theorem of Pythagoras applies: $c^2 = a^2 + b^2$.
- In order to use Pythagoras, the triangle must be right-angled **and** you must know two out of the three sides.

1. Find the value of the missing side in each triangle below.

   a)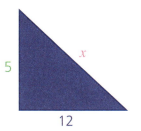

   Here, we are trying to find the value of the hypotenuse (x).
   We know two sides and want the third, so we use the theorem of Pythagoras:
   $c^2 = a^2 + b^2$
   $x^2 = 5^2 + 12^2$
   $x^2 =$

   b)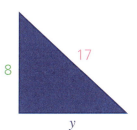

   Here, we are trying to find the value of the adjacent side (y).
   We know two sides and want the third, so we use the theorem of Pythagoras:
   $c^2 = a^2 + b^2$
   This time, we know the hypotenuse and another side. Therefore:
   $(17)^2 = (8)^2 + (y)^2$
   $(17)^2 - (8)^2 = y^2$
   $ = y^2$

CHAPTER 10: TRIGONOMETRY

Section C: Using Trigonometric Ratios to Find Missing Sides in a Right-Angled Triangle

- In Questions 2 and 3, we will not be able to use the theorem of Pythagoras to find the value of the missing side. To use Pythagoras, we would need to know two sides out of the three.

- Instead we will have to use one of the trigonometric ratios to find the missing value.

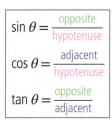

$\sin \theta = \dfrac{\text{opposite}}{\text{hypotenuse}}$

$\cos \theta = \dfrac{\text{adjacent}}{\text{hypotenuse}}$

$\tan \theta = \dfrac{\text{opposite}}{\text{adjacent}}$

Worked example

In this triangle, calculate the length of the missing side, to the nearest centimetre.

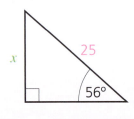

x = opposite   25 = hypotenuse

So the trigonometric ratio we will use is sin.

$\sin \theta = \dfrac{\text{opposite}}{\text{hypotenuse}}$

$\sin 56° = \dfrac{x}{25}$

$(25)\sin 56° = \dfrac{x}{25}(25)$

$20{\cdot}7259 = x$

$21 \text{ cm} = x$ (to the nearest cm)

To create a linear equation, we must multiply both sides of the equals by 25 (the denominator of the fraction).

2. In the following triangle, calculate the value of the missing side, to the nearest centimetre.

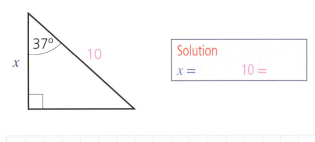

Solution
$x =$ $10 =$

Step 1: Label the sides of the triangle as opposite, adjacent and hypotenuse.
Step 2: Choose the trigonometric ratio that will allow you to solve the unknown.

146 SKILLS FOR EXAM SUCCESS MATHS

Worked example

Calculate the length of the missing side in this triangle, correct to 2 decimal places.

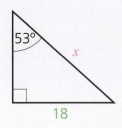

Solution

x = hypotenuse, 18 = opposite, so the trigonometric ratio we need to solve for x is sin.

$\sin \theta = \dfrac{\text{opposite}}{\text{hypotenuse}}$

$\sin 53° = \dfrac{18}{x}$

$x \sin 53° = \dfrac{18}{x}(x)$

$x = \dfrac{18}{\sin 53°}$

$x = 22\cdot538 = 22\cdot54$ units

To form a linear equation, we must multiply both sides of the equals by x (the denominator of the fraction).

Exam Hint

In the exam, there will be marks for each correct step that you make. So even if you can only do one part of the question, try that much. You never know what Low Partial Credit will be allocated for. Some marks are better than no marks!

3. In the following triangles, calculate the value of the missing side, correct to 2 decimal places.

a)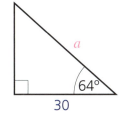

Solution
30 = adjacent a = hypotenuse

CHAPTER 10: TRIGONOMETRY   **147**

b)

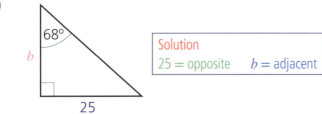

Solution
25 = opposite   b = adjacent

Section D: Using Trigonometric Ratios to Find Missing Angles

4. If $\cos \theta = \frac{12}{13}$, find, as fractions, the value of $\sin \theta$ and $\tan \theta$.

Step 1: $\cos \theta = \frac{12}{13}$

$\cos \theta = \frac{\text{adjacent}}{\text{hypotenuse}}$. This tells us that 12 = adjacent and 13 = hypotenuse.

In order to find $\sin \theta$ and $\tan \theta$, we need to know the remaining side, the opposite.

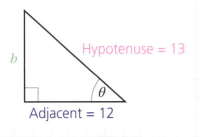

Step 2: To find the remaining side we use Pythagoras:

$c^2 = a^2 + b^2$
$13^2 = 12^2 + b^2$

Step 3: We now know the opposite, the adjacent and the hypotenuse. Using this, find the value of $\sin \theta$ and $\tan \theta$ as fractions.

$\sin \theta = \frac{\text{opposite}}{\text{hypotenuse}} = \underline{}$ $\tan \theta = \frac{\text{opposite}}{\text{adjacent}} = \underline{}$

SKILLS FOR EXAM SUCCESS MATHS

5. Find the value of A, to the nearest degree, in the following.

Worked example

cos A = 0·6990

A = cos⁻¹(0·6990)

A = 45·65°

A = 46°

Here you are given the cosine of the angle A. You want to know what the angle A is. To find angle A, we use the inverse function, here cos⁻¹. As all calculators are different, make sure you know how to get inverse trig functions on yours.

a) sin A = 0·9653

b) tan A = 0·4270

6. In the following triangles, calculate the missing angle, correct to 2 decimal places.

Worked example

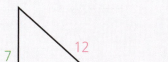

7 = opposite, 12 = hypotenuse, so the Trig ratio we need to use here is sin.

$\sin A = \dfrac{\text{opposite}}{\text{hypotenuse}}$

$\sin A = \dfrac{7}{12}$

$A = \sin^{-1} \dfrac{7}{12}$

A = 35·6853

A = 35·69°

Label the sides of the triangle as opposite, adjacent and hypotenuse. Choose the trigonometric ratio that will allow you to solve the unknown.

a)

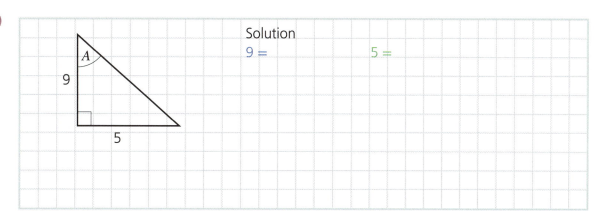

Solution

9 =   5 =

CHAPTER 10: TRIGONOMETRY   149

b)

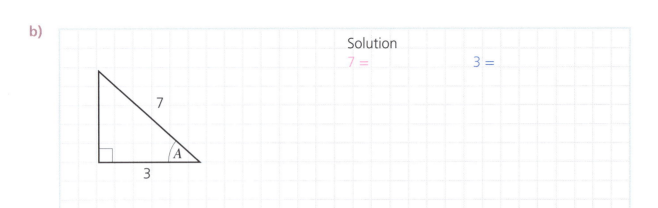

Solution

7 =

3 =

Section E: Using Trigonometry to Solve Real-Life Problems

Exam Hint

When working on any Trig question in the exam, always make sure you write $\sin \theta = \frac{opposite}{hypotenuse}$, $\cos \theta = \frac{adjacent}{hypotenuse}$, $\tan \theta = \frac{opposite}{adjacent}$ and the theorem of Pythagoras ($c^2 = a^2 + b^2$) in your answer booklet, even if you are not sure how to answer the question. You never know what Low Partial Credit may be awarded for in the exam.

It is not enough to write SOHCAHTOA or any of the other memory aids. You must write the Trig ratios in the format shown above.

7. The picture on the right shows a house. Part of the roof of this house is shown in the diagram below.

    AB is perpendicular to DC.

    $|AD| = 11·8$ m and $|DB| = 2$ m

    $|\angle CAD| = 20°$ and $|\angle DBC| = x°$

    Note: $\angle ACB$ is **not** a right angle.

    Use trigonometry to work out the value of x.

    Give your answer correct to the nearest whole number.

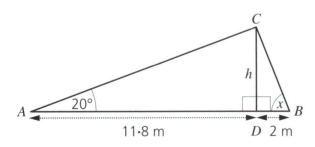

1. Using trigonometric ratios, look at the larger triangle and find the value of h.
2. Using trigonometric ratios, use h and $|DB|$ to find the value for x.

Step 1:

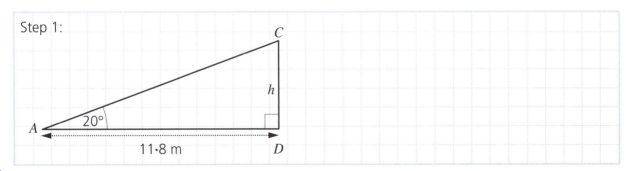

150 SKILLS FOR EXAM SUCCESS MATHS

Step 2:

8. a) Use trigonometry to work out the value of x in the triangle below.

    Give your answer correct to 2 decimal places.

    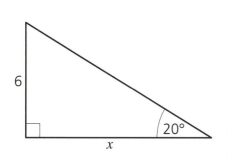

    b) A different right-angled triangle is shown on the right.

    It has sides of length a, b and c.

    One of the angles is Y.

    In this triangle, $a + b > c$.

    Use this to prove that $\cos Y + \sin Y > 1$.

    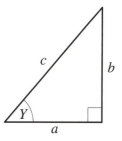

Exam Hint

In the exam, there will be marks for each correct step that you make. So even if you can only do one part of the question, try that much. You never know what Low Partial Credit will be allocated for. Some marks are better than no marks!

9. A tree 32 m high casts a shadow 63 m long. Calculate θ, the angle of elevation of the sun. Give your answer correct to 3 significant figures.

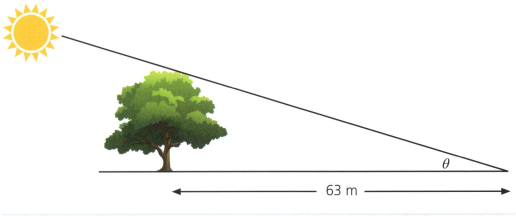

Exam Hint

Even if you are unsure how to give the answer in the format that is required, do as much of the question as you can. There are marks available for each step along the way. You may only lose one or two marks for not giving your final answer in the correct format.

10. The top of Fastnet Lighthouse, F, is 49 m above sea level.

The angle of elevation to the top of the lighthouse from a ship, S, is 1·2°, as shown in the diagram, which is not to scale.

Find the horizontal distance, marked d, from the ship to the base of the lighthouse.

Give your answer in kilometres, correct to 2 decimal places.

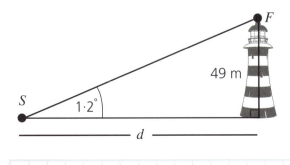

Remember: Marks will be deducted for failing to round your answer or for rounding incorrectly.

152 SKILLS FOR EXAM SUCCESS MATHS

11. A vertical mobile phone mast, *DC*, of height *h* m, is secured to two cables: *AC* of length *x* m, and *BC* of length 51·8 m, as shown in the diagram.

The angle of elevation to the top of the mast from *A* is 30° and from *B* is 45°.

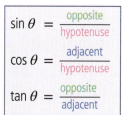

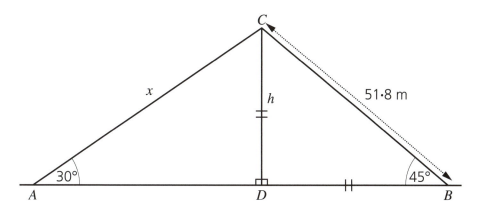

a) Find the value of *h*, the height of the phone mast.

b) Find the value of *x*, the length of the cable |AC|.

c) Hence, or otherwise, find the length of |AD|.

CHAPTER 10: TRIGONOMETRY

Exam-Style Questions

Exam Hints

1. Attempt every part of every question. You cannot get marks for a blank space!
2. Highlight key words in the question. Make sure you understand what you are being asked to do.
3. Write the relevant formula. The correct formula with some relevant substitution will get you marks
4. If you are asked to give your answer in a particular format and are unsure of how to do this, do as much as you can.
5. Marks are not just given for the correct answer, they are also available for work completed along the way.

Question 1

Linked Question: Trigonometry, Distance/Time/Speed, Graphing Functions, Algebra

a) When a hurler stands at the 45 m line, the angle of elevation to a target on a net behind the goals is 15°.

   Calculate the height of the target above the ground, to 2 decimal places.

   Exam Hint
   Marks will be deducted for failing to round your answer or for rounding incorrectly.

b) The hurler now stands at the 60 m line. Calculate the angle of elevation now from the hurler to the target, correct to 2 decimal places.

c) The hurler hits the sliotar from the 60 m line. Three seconds later the sliotar hits the target. Calculate the speed of the ball. Give your answer in kilometres per hour (km/hr).

   Speed = Distance/Time
   1 km = 1000 m
   1 hour = 3600 seconds

d) During a match, Keith hits the ball with his hurley. The height of the ball can be modelled by the following quadratic function:

$$h = -2t^2 + 5t + 1\cdot2$$

where h is the height of the ball in metres and t is seconds after being hit, where $t \in \mathbb{R}$.

i. How high, in metres, was the ball when it was hit (when $t = 0$)?

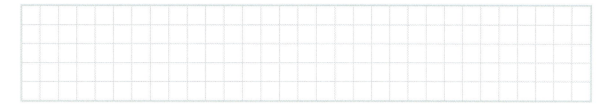

ii. The ball was caught after 2·4 seconds.

How high, in metres, was the ball when it was caught?

Hint: $t = 2\cdot4$

e) When the ball passed over the halfway line, it was at a height of 3·2 metres and its height was decreasing.

How many seconds after it was hit did the ball pass over the halfway line?

Remember: $h = -2t^2 + 5t + 1\cdot2$

Hint:
$$x = \frac{-b \pm \sqrt{b^2 - 4ac}}{2a}$$

$-2t^2 + 5t + 1\cdot2 = 3\cdot2$

$-2t^2 + 5t + 1\cdot2 - 3\cdot2 = 0$

$-2t^2 + 5t - 2 = 0$

Rearrange to form a quadratic equation.

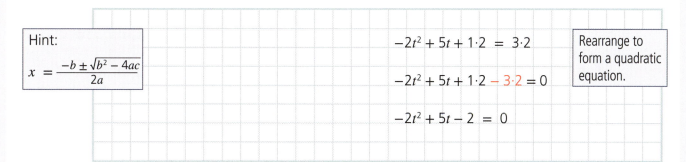

CHAPTER 10: TRIGONOMETRY

f) Draw a graph of the function $-2t^2 + 5t + 1.2$ in the domain $0 \leq t \leq 2.5$.

t	$-2t^2 + 5t + 1.2$	h	Point
0.5			
1			
1.5			
2			
2.5			

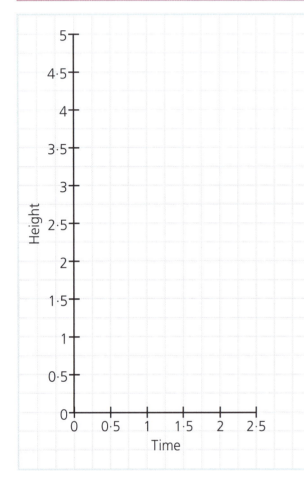

Remember!

When graphing functions:
1. Always use a pencil.
2. Label the *x*- and the *y*-axes.
3. Make sure the scale along the *x*-axis is constant.
4. Make sure the scale along the *y*-axis is constant.
5. Remember to plot the *x* co-ordinate first and then the *y*. For example, for the point (−2, −7), go to −2 on the *x*-axis first, then −7 on the *y*-axis.
6. Plot the points clearly and accurately.
7. Finally, join your points together. There are marks available for plotting the points correctly **and** joining them together.

g) Using your graph, answer the following questions.

   i. What is the maximum height reached by the ball?

   ii. At approximately what time will the ball hit the ground?

   iii. Draw the axis of symmetry on your graph.

SKILLS FOR EXAM SUCCESS MATHS

Question 2 – 2022

The diagram below shows two vertical buildings, A and B (diagram not to scale).

Mary stands at the top of Building A. She is 220 m above the ground.

She wants to work out the distances marked y and z in the diagram – that is, the distance from the top of Building A to the bottom of Building B, and the height of Building B, respectively.

Mary measures the two angles that are marked 35° and 20° in the diagram, to the bottom of Building B and the top of Building B, respectively. The broken line is horizontal.

Exam Hint

I recommend that you write out each step clearly so that if you do make a mistake, the examiner will be able to pinpoint exactly where your error is and deduct the minimum amount of marks.

a) Work out the size of the angle C.

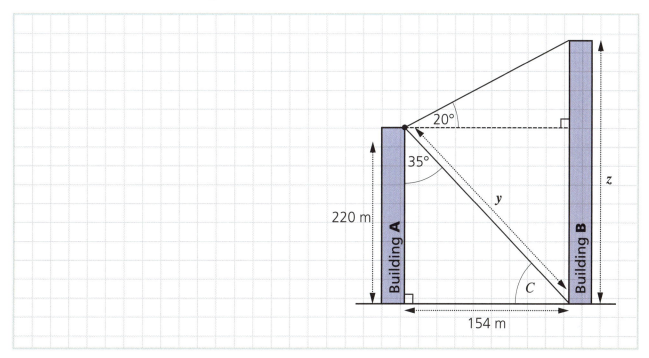

b) Mary works out that the horizontal distance between the two buildings is 154 m, correct to the nearest metre, as shown.

Use the theorem of Pythagoras to work out the distance marked y on the diagram. Give your answer correct to the nearest metre.

c) Use trigonometry to work out the value of z, the height of Building B. Give your answer correct to the nearest metre.

Remember!

Remember: Marks will be deducted for failing to round your answer or rounding incorrectly.

CHAPTER 10: TRIGONOMETRY

Question 3

Linked Question: Trigonometry, Geometry, Numbers

a) The triangle *XYZ* has the following sides: |*XZ*| = 7 cm and |*ZY*| = |*XY*| = 5 cm.

Construct the triangle, showing all construction lines clearly.

Exam Hint
All construction lines must be visible to the examiner for full marks.

b) i. What type of triangle is the triangle *XYZ*?

Equilateral **Isosceles** **Scalene**
☐ ☐ ☐

ii. Justify your answer.

Exam Hint
There are marks available for **both** ticking the correct box **and** for justifying your answer. Even if you find it difficult to give a reason for your answer, still tick the box you think is correct for some marks.

c) Construct the perpendicular bisector of |*XZ*|.

Extend the line so that it passes through the point *Y*.

Where the bisector cuts the line |*XZ*|, label this point *S*.

Show all construction lines clearly.

d) Using trigonometry, find the length of the side |*SY*|.
Give your answer correct to 1 decimal place.

Exam Hint
Even if you are unsure how to give the answer in the format that is required, do as much of the question as you can! You may only lose one or two marks for not giving your final answer in the correct format.

158 SKILLS FOR EXAM SUCCESS MATHS

e) Using trigonometry, find the value of ∠SZY correct to 2 decimal places.

f) Prove that triangle SZY is congruent to triangle XSY.

Conditions for congruency:
SSS
SAS
ASA
RHS

Question 4

Linked Question: Co-ordinate Geometry of the Line, Trigonometry

a) i. On the co-ordinate plane, plot the following points:

   $A = (5, 2)$ $B = (1, 2)$ $C = (1, 5)$

   Join the points together to form a triangle.

   Label ∠BAC as A.

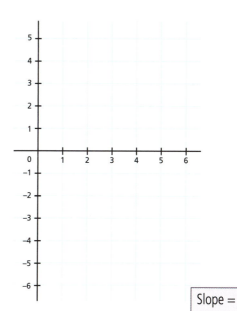

ii. Find the slope of the line |AC|.

Slope = $\dfrac{y_2 - y_1}{x_2 - x_1}$

b) Find the length of the following sides.

   i. |BC| ii. |AB|

c) i. Hence, or otherwise, find tan A.

$$\tan A = \frac{\text{opposite}}{\text{adjacent}}$$

ii. Comment on the relationship between the slope of the line |AC| and tan A.

d) On the graph you created in part (a), find the image of the triangle ABC in the x-axis. Label the image A'B'C'.

e) Prove that the slope of the line $|A'C'| = \tan A'$

Step 1: Find the slope of A'C' using the formula:
$$\frac{y_2 - y_1}{x_2 - x_1}$$
Step 2: Find the length of $|A'B'|$ and $|B'C'|$.

Step 3: Find tan A'.

$$\tan A = \frac{\text{opposite}}{\text{adjacent}}$$

Summary review:

- You need to remember:
  - $\sin \theta = \frac{\text{opposite}}{\text{hypotenuse}}$ $\cos \theta = \frac{\text{adjacent}}{\text{hypotenuse}}$ $\tan \theta = \frac{\text{opposite}}{\text{adjacent}}$

  These are not in the *Formulae and Tables* booklet and must be remembered.

  - The three angles in a triangle add to 180°.
  - The theorem of Pythagoras: $c^2 = a^2 + b^2$.

- You need to be familiar with the settings on your own calculator to evaluate sin, tan, cos, $\sin^{-1}$, $\tan^{-1}$ and $\cos^{-1}$.

Scan this code to access a video with worked solutions for every question.

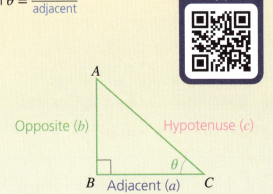

160 SKILLS FOR EXAM SUCCESS MATHS

Chapter 11: Geometry 1

The topics and skills I will work on are:

Terminology – theorem, proof, axiom, corollary, converse, implies ☐

Different types of angles and lines ☐
- **Perpendicular and parallel lines** ☐
- **Corresponding, alternate, vertically opposite and interior angles with parallel lines** ☐
- **Straight, acute, obtuse, right and reflex angles** ☐

Properties of triangles ☐

Properties of quadrilaterals ☐

Proofs ☐

Properties of angles within circles ☐

Section A: Terminology

1. Complete the following definitions by adding explanations and examples for each.

   a) A **theorem** is: _____

   Example: Vertically opposite angles are equal in measure.

   b) A **proof** is: _____

   Example: To prove that vertically opposite angles are equal in measure.

   ($|\angle 1| = |\angle 2|$)

   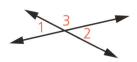

Statement	Reason
$|\angle 1| + |\angle 3| = 180°$	Straight angle = 180°
$|\angle 2| + |\angle 3| = 180°$	Straight angle = 180°
$|\angle 1| + |\angle 3| = |\angle 2| + |\angle 3|$	Subtract $|\angle 3|$ from both sides
$|\angle 1| = |\angle 2|$	

c) An **axiom** is: _____

   Example: _____

d) A **corollary** is: _____

   Example: _____

> **Remember!**
> **Implies** (⇒): is used in a proof when we can write a fact we have proven by our previous statements.

e) A **converse** is: _____

   Example: _____

Section B: Different Types of Angles and Lines

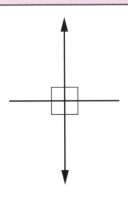

Perpendicular lines:
- Lines that intersect at 90°.
- The slope of line 1 is m and the slope of line 2 is $-\frac{1}{m}$.
- Symbol: ⊥

Parallel lines:
- Lines that never intersect and are always the same distance apart.
- Slope of both lines is the same.
- Symbol: ∥

Angles:
- You will need to be familiar with the following types of angle:

Corresponding angles

Alternate angles

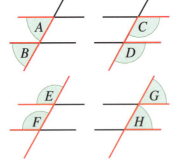

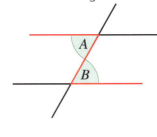

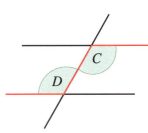

Vertically opposite angles

Interior angles

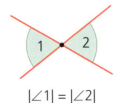

$|\angle 1| = |\angle 2|$

Interior angles between two parallel lines add up to 180°

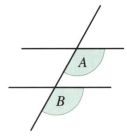

$|\angle A| = |\angle B| = 180°$

SKILLS FOR EXAM SUCCESS MATHS

2. In the diagram below, the line *l* is parallel to the line *k*.

   The angles *A*, *B*, *C*, *D*, *E*, *F*, *G* and *H* are marked on the diagram.

   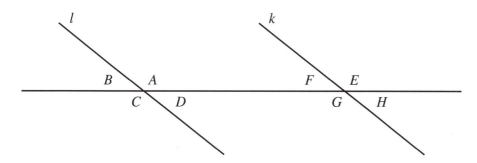

   a) Write down a pair of angles that are **vertically opposite**.

   _____ and _____

   b) Write down a pair of angles that are **corresponding**.

   _____ and _____

   c) Write down a pair of angles that are **alternate**.

   _____ and _____

   d) Write down a pair of **interior angles**.

   _____ and _____

   > For alternate, corresponding and interior angles, the lines involved must be parallel and the angles are formed by a transversal line that passes through both parallel lines.
   >
   >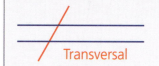
   > Transversal

   e) i. Given that |∠A| = 137°, find the measure of each of the angles *B*, *C*, *D*, *E*, *F*, *G* and *H*.

   ii. Give a reason for each answer.

Angle	Measure	Reason
B		
C		
D		
E		
F		
G		
H		

3. a) Four angles are shown below. Write in the space below each diagram whether the angle is straight, acute, obtuse, right or reflex.

   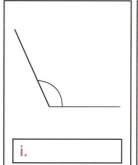 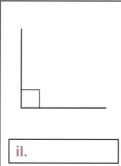  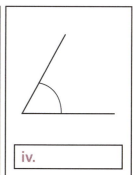

   i. _____   ii. _____   iii. _____   iv. _____

b) In the diagram below, $l_1 \parallel l_2$. Write the measure of each angle shown by an empty box into the diagram, without using a protractor.

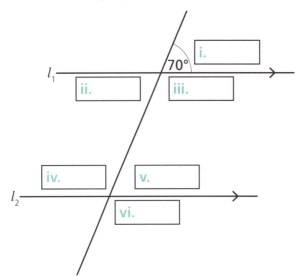

Hint: There are two parallel lines cut by a transversal, so you should be looking out for alternate, corresponding and interior angles.

4. Six angles are formed by three straight lines, as shown below. Find the missing angles and write them in the table.

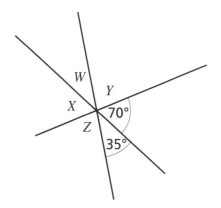

W	
X	
Y	
Z	

Hint: Vertically opposite angles are equal in measure **and** there are 180° in a straight angle.

5. If l_1, l_2 and l_3 are parallel lines, find the measure of the angles A, B and C.

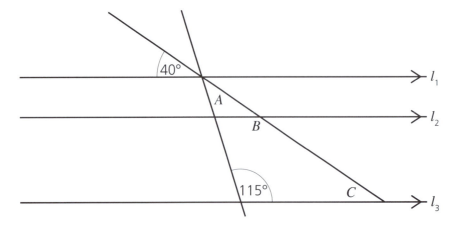

Remember: Alternate, corresponding and vertically opposite angles. Sometimes you may have to find the measure of an angle that you do not need in your answer before you can find the measure of the required angle.

164 SKILLS FOR EXAM SUCCESS MATHS

Angle A:

Angle B:

Angle C:

Section C: Properties of Triangles

Equilateral	Isosceles	Scalene
All sides are the same length.	Two sides are the same length.	No sides are the same length.
All angles are the same size (60°).	Two angles are the same size.	No angles are the same size.

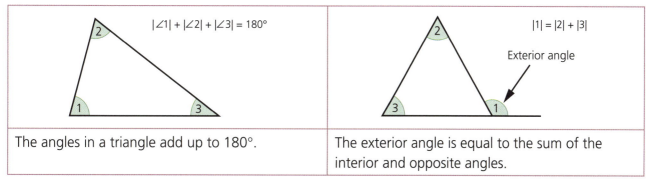

The angles in a triangle add up to 180°.	The exterior angle is equal to the sum of the interior and opposite angles.

- In any triangle, the largest angle is opposite the longest side.
- The smallest angle is opposite the shortest side.
- The converse of these statements is also true.

Remember!
Remember the theorem of Pythagoras for a right-angled triangle: $c^2 = a^2 + b^2$.

CHAPTER 11: GEOMETRY 1

Section D: Properties of Quadrilaterals

Type of quadrilateral	Sides	Parallel sides	Angles	Diagonals
Parallelogram	Opposite sides are equal	Opposite sides are parallel	Opposite angles are equal	Bisect each other
Rhombus	Four equal sides	Opposite sides are parallel	Opposite angles are equal	Bisect each other – angle of 90° formed
Rectangle	Opposite sides are equal	Opposite sides are parallel	All angles the same size (90°)	Bisect each other
Square	Four equal sides	Opposite sides are parallel	All angles the same size (90°)	Bisect each other – angle of 90° formed

6. If $l_1 \parallel l_2$, find the sizes of the angles A, B and C in the following diagram.

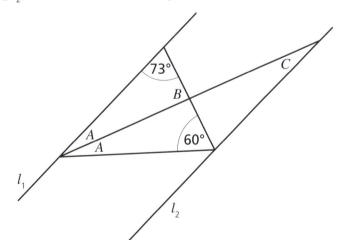

Note: It is OK to have your answer as a decimal here.

Hint:
3 angles in a triangle add up to 180°
Alternate angles

Angle	Workings	Answer
A		
B		
C		

7. The diagram below shows a parallelogram. One of the sides has been extended.

   The sizes of some of the angles are marked.

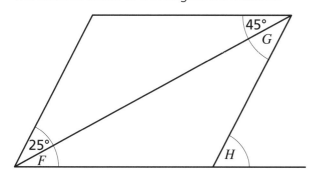

Hint:
Alternate angles
The exterior angle in a triangle is equal to the sum of the interior and opposite angles.

Find the size of the angles *F*, *G* and *H* **without** measuring.

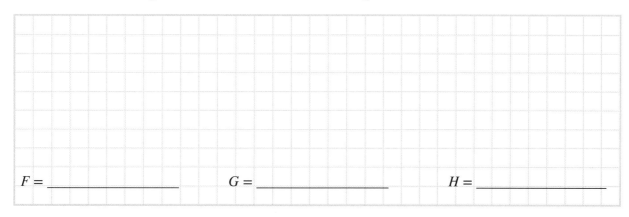

F = _____   G = _____   H = _____

8. The diagram below shows a parallelogram and its two diagonals.

   Some of the angles in the diagram are marked.

   Write down the size of the angle *K* and the size of the angle *L*.

   Give a reason for each answer.

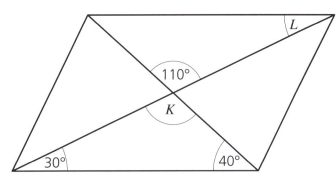

Hint:
Vertically opposite angles
Alternate angles

	Angle	Reason
K		
L		

Section E: Proofs

- A proof is a step-by-step explanation that uses axioms and previously proven theorems to show that another theorem is correct.

- Each statement made must be justified with a reason using evidence from the diagram or from other theorems or axioms.

9. Prove that $X + Y + Z = 360°$ by completing the table on the next page.

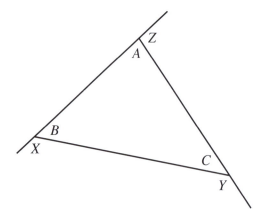

For full marks, you need to give a reason for each statement you make. You cannot make any assumptions when carrying out a proof. You can only work with facts.

Statement	Reason
$\lvert\angle A\rvert + \lvert\angle B\rvert + \lvert\angle C\rvert = 180°$	
$\lvert\angle A\rvert + \lvert\angle Z\rvert = 180°$	
Similarly _____ + _____ = 180° **AND** _____ + _____ = 180°	
Therefore $Z + A + B + X + C + Y =$ _____°	
$A + B + C =$ _____° $Z + X + Y +$ _____° $= 540°$	
Therefore $Z + X + Y =$	

10. Prove that $\lvert\angle 4\rvert = \lvert\angle 1\rvert + \lvert\angle 2\rvert$

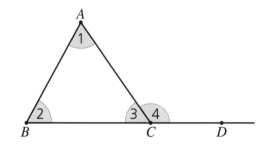

Statement	Reason
$\lvert\angle 1\rvert + \lvert\angle 2\rvert + \lvert\angle 3\rvert =$	
$\lvert\angle 3\rvert + \lvert\angle 4\rvert = 180°$	
Therefore $\lvert\angle 1\rvert + \lvert\angle 2\rvert + \lvert\angle 3\rvert = \lvert\angle 3\rvert + \lvert\angle 4\rvert$	
So $\lvert\angle 1\rvert + \lvert\angle 2\rvert =$ _____	

11. In the square ABCD, prove that ∠3 = 45°.

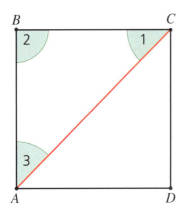

Statement	Reason
$\lvert\angle 2\rvert =$ _____	The four angles in a square are 90°
$\lvert AB\rvert = \lvert BC\rvert$	
$\angle$ _____ $= \angle$ _____	In an isosceles triangle, the angles opposite the equal sides are equal in measure
$\lvert\angle 1\rvert + \lvert\angle 2\rvert + \lvert\angle 3\rvert = 180°$	
$\lvert\angle 1\rvert + 90 + \lvert\angle 3\rvert = 180°$	
$\lvert\angle 1\rvert + \lvert\angle 3\rvert =$ _____	
Therefore ∠3 = 45°	

Section F: Properties of Angles Within Circles

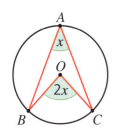

The angle at the centre of a circle is twice the angle at the circumference, if they are standing on the same arc.

∠BOC = 2(∠BAC)

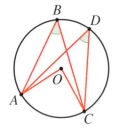

Angles standing on the same arc are equal in measure.

∠ABC = ∠ADC

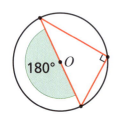

The angle in the semi-circle is a right angle (90°).

The diameter of the circle is the hypotenuse of the right-angled triangle.

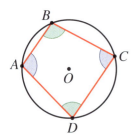

A cyclic quadrilateral is a quadrilateral within a circle where all four corners touch the circumference of the circle.

In a cyclic quadrilateral, opposite angles add up to 180°.

So here, $\lvert\angle A\rvert + \lvert\angle C\rvert = 180$ and $\lvert\angle B\rvert + \lvert\angle D\rvert = 180$.

$\lvert\angle A\rvert + \lvert\angle B\rvert + \lvert\angle C\rvert + \lvert\angle D\rvert = 360°$

12. Prove that ∠A = ∠B.

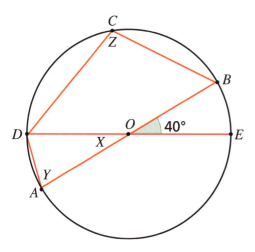

Statement	Reason
$|\angle C| = 2|\angle A|$	
$|\angle C| = 2|\angle B|$	
Therefore $2|\angle A| = 2|\angle B|$ So $|\angle A| = |\angle B|$	

13. The diagram below shows a circle with centre O. The five points A, B, C, D and E are on the circle.

    [AB] and [DE] are diameters of the circle and $|\angle BOE| = 40°$.

    The angles X, Y and Z are labelled.

a) Calculate the value of X.

   Give a reason for your answer.

CHAPTER 11: GEOMETRY 1 | 171

b) Calculate the value of Y. Give a reason for your answer.

Hint:
|DO| = |AO| as they are both radii of the circle. This makes the triangle DOA an isosceles triangle.

c) Find the value of Z. Show all your workings.

Hint:
40° + ∠O = 180° as they form a right angle.
and
Opposite angles in a cyclic quadrilateral add to 180°.

14. The points A, B, C and D are shown on the diagram. They are all on the circle k.

|AB| = |AD| and |BC| = |DC|, as shown.

The size of one of the angles is marked.

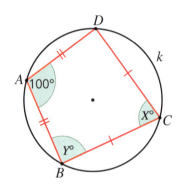

a) Find the value of X.

Give a reason for your answer.

Hint: Opposite angles in a cyclic quadrilateral add to 180°.

b) Calculate the value of Y. Show all your workings.

Hint: Draw the line [BD] on your diagram. Doing this creates the triangle ABD (which is an isosceles triangle) and the triangle BCD (which is also an isosceles triangle). Apply your knowledge about isosceles triangles.

SKILLS FOR EXAM SUCCESS MATHS

Exam-Style Questions

Exam Hints

1. Attempt every part of every question. You cannot get marks for a blank space!
2. Highlight key words in the question. Make sure you understand what you are being asked to do.
3. If you are asked to give your answer in a particular format and are unsure of how to do this, do as much as you can.
4. Marks are not just given for the correct answer, they are also available for work completed along the way.
5. When working through geometry questions, it is sometimes necessary to find the measure of an angle that you do not need in your answer before you can find the measure of the required angle.

Question 1

Linked Question – Geometry, Algebra

In the diagram to the right: $|\angle CAB| = 80°$, $|\angle DCE| = 60°$, $|\angle ABC| = (x + y)°$ and $|\angle BCA| = (3x + y)°$, where $x, y \in \mathbb{N}$.

Find the value of x and the value of y.

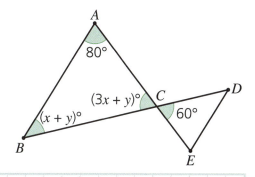

Firstly, we have to form two equations in x and y.

Equation 1: $x + y + 3x + y + 80 = 180$
$4x + 2y = 180 - 80$
$4x + 2y = 100$

The three angles in a triangle add to 180°

Equation 2: $3x + y = 60$ Vertically opposite angles are equal in measure

Now use simultaneous equations to solve for x and y.
$4x + 2y = 100$
$3x + y = 60$

Steps in solving simultaneous equations:
1. Use the elimination method to cancel one of the variables.
2. Find the value of one variable.
3. Use substitution to find the value of the other variable.

CHAPTER 11: GEOMETRY 1 **173**

Question 2

a) The diagram below shows a parallelogram and one exterior angle. Find the value of a and the value of b.

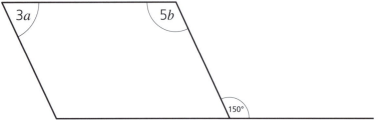

Hint:
Step 1: $3a + 150 = 180°$
Step 2: Opposite angles in a parallelogram are equal in measure.
Step 3: $5b$ and 150 are alternate.

b) The circle c is shown in the diagram below (not to scale). Its centre is at the point O. The points A, B and D lie on the circle, and $[AB]$ is a diameter of the circle.

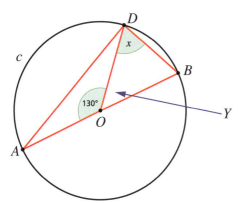

i. Write down $|\angle ADB|$, the size of the total angle at the point D.

Hint: Angle in the semi-circle.

ii. $|\angle AOD| = 130°$. Work out the size of the angle marked X in the diagram.

Step 1: Find the angle Y. Sometimes you may have to find the measure of an angle that you do not need in your answer before you can find the measure of the required angle.
Step 2: $|OD| = |OB|$ as they are both radii.
Step 3: Look at the triangle DOB as an isosceles triangle.

SKILLS FOR EXAM SUCCESS MATHS

c) Use the diagram below to prove that the three angles in a triangle add to 180°.

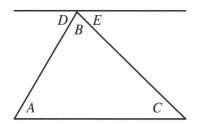

Statement	Reason
$\|\angle D\| + \|\angle E\| + \|\angle B\| = 180°$	
$\|\angle A\| = \|\angle D\|$	
$\|\angle \text{_____}\| = \|\angle \text{_____}\|$	Alternate angles
Therefore ∠A + ∠B + ∠C = 180°	

d) The diagram below shows a parallelogram, with one side extended.

Use the data on the diagram to find the value of X, Y and Z.

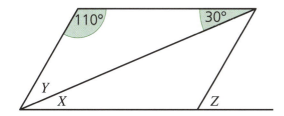

Remember!
The three angles in a triangle add to 180°.
Alternate angles are equal in measure.
The exterior angle is equal to the sum of the interior and opposite angles.

Angle	Reason
X =	
Y =	
Z =	

Exam Hint
You must give a reason for each answer in order to get full marks. However if you struggle to give your reason, at least give the measure of the angle. Some marks are better than no marks.

CHAPTER 11: GEOMETRY 1

Question 3

Linked Question – Geometry, Trigonometry, Arithmetic

A vertical phone mast, [DC], of height h m, is secured with two cables: [AC] of length x m, and [BC] of length 51·8 m, as shown in the diagram.

The angle of elevation to the top of the mast from A is 30° and from B is 45°.

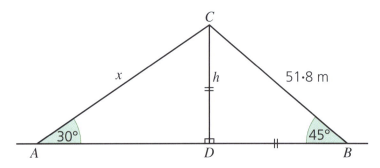

a) Explain why $|\angle BCA| = 105°$.

Hint: Look at the triangle ABC.

b) Using trigonometric ratios, find the value of h, the height of the vertical phone mast correct to 1 decimal place.

$\sin \theta = \dfrac{\text{opposite}}{\text{hypotenuse}}$

$\cos \theta = \dfrac{\text{adjacent}}{\text{hypotenuse}}$

$\tan \theta = \dfrac{\text{opposite}}{\text{adjacent}}$

c) Hence, or otherwise, find the value of x, the length of cable.

d) The cable required to secure the mast costs €25 per metre. The mast itself costs €580 per metre.

VAT at 23% is then added in each case.

Calculate the total cost of the cable and mast after VAT is included.

Exam Hint

As you can see, the answers from part (c) are used in part (d). Even if you got the wrong answer in part (c), if you use that answer in part (d) correctly, you will get full marks!

176 SKILLS FOR EXAM SUCCESS MATHS

Step 1: Add your value for x and 51·8 m to find the total length of cable required.

Step 2: Multiply your answer by €25 to find the cost of the cable.

Step 3: Add €580 to find the total cost of the materials.

Step 4: Calculate 23% VAT and add to find total bill.

Summary review:

You need to remember the following:

Theorem 1: Vertically opposite angles are equal in measure.

Theorem 2: In an isosceles triangle, the angles opposite the equal sides are equal in measure.

Theorem 3: If a transversal makes equal alternate angles on two lines then the lines are parallel.

Theorem 4: The angles in a triangle add to 180°.

Theorem 5: Two lines are parallel if and only if, for any transversal, the corresponding angles are equal.

Theorem 6: Each exterior angle of a triangle is equal to the sum of the interior opposite angles.

Theorem 9: In a parallelogram, opposite sides are equal and opposite angles are equal.

Theorem 10: The diagonals of a parallelogram bisect each other.

Theorem 14: In a right-angle triangle, the square of the hypotenuse is equal to the sum of the squares of the other two sides.

Theorem 15: If the square of one side of a triangle is equal to the sum of the squares of the other two sides, then the angle opposite the first side is a right angle.

Theorem 19: The angle at the centre of a circle standing on a given arc is twice the angle at any of the circle standing on the same arc.

Corollary 2: All angles at points of the circle, standing on the same arc, are equal.

Corollary 3: Each angle in a semi-circle is a right angle.

Corollary 4: If the angle standing on a chord at some point of the circle is a right angle, then the chord is a diameter.

Corollary 5: In a cyclic quadrilateral, opposite angles sum to 180°.

Scan this code to access a video with worked solutions for every question.

CHAPTER 11: GEOMETRY 1

Chapter 12: Applied Measure

> The topics and skills I will work on are:
>
> **Area and perimeter of rectangles, squares, parallelograms, circles and triangles**
> - Circumference and area of circles
> - Area and length of an arc
> - Finding the missing dimension when given the area/perimeter/circumference
>
> **Volume, surface area and nets of rectangular solids and cubes**
>
> **Prisms**
>
> **Cylinders, spheres and hemispheres**
>
> **Compound volumes**
>
> **Equal volumes**

Note: In this section of maths, there are a few points you need to be extra aware of:

- Units in answers. If the question asks you to give your answer in m³, for example, no additional marks are given for including m³ in your answer. If there is no reference to units in the question, marks will be deducted for not correctly stating the units in your answer.

- Make sure your answers are in the correct format, e.g. in terms of π or to the correct number of decimal places.

- I would advise putting the value you are substituting into a formula in brackets, as this will help to reduce any mistakes with signs.

- Many of the formulae you require are on pages 8, 9 and 10 of the *Formulae and Tables* booklet.

Section A: Area and Perimeter of Rectangles, Squares, Parallelograms, Circles and Triangles

- Some of the formulae for calculating area are on pages 8 and 9 of the *Formulae and Tables* booklet. You need to remember the formulae for calculating the area and perimeter of a square and a rectangle and the perimeter of a parallelogram as these are not in the tables.

- Area = the space enclosed by the boundary lines.

- Units for area are units².

- Perimeter = the distance around the edge of a shape.

Remember!
When two letters are written beside each other in a formula, this means that the values must be multiplied together. For example:
Area of a triangle = $\frac{1}{2}bh$
means $\frac{1}{2} \times b \times h$.

SKILLS FOR EXAM SUCCESS MATHS

1. Calculate the area and the perimeter of the following shapes.

Worked example

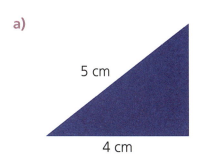

Area = lw
  $(8) \times (5)$
  40 cm^2

Perimeter = $2l + 2w$
  $2(8) + 2(5)$
  26 cm

Remember to write the relevant formula so that if you make a mistake, you will still receive some marks. Don't forget the units.

a)

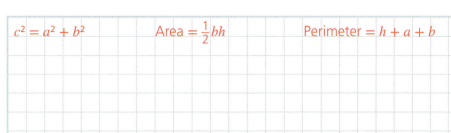

$c^2 = a^2 + b^2$ Area = $\frac{1}{2}bh$ Perimeter = $h + a + b$

In order to find the area of a triangle, you need to know the base and the height. Here we do not know the height but we can use the theorem of Pythagoras to find it.

b)

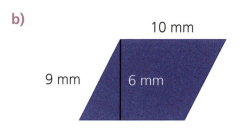

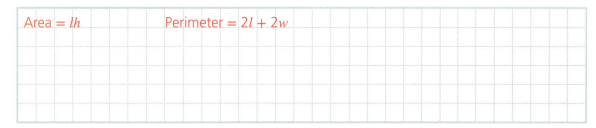

Area = lh Perimeter = $2l + 2w$

CHAPTER 12: APPLIED MEASURE   179

c) The circle has a diameter of 10 cm.

Give your answers in terms of π.

Area = πr^2 Circumference = $2\pi r$

2. Calculate the area of the square in terms of a, given that the area of the right-angled triangle is $12a^2$.

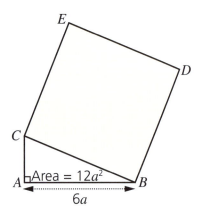

Worked example

There are three steps to this question.

Step 1: We must find the height of the triangle.

Area = $\frac{1}{2}bh$

$12a^2 = \frac{1}{2}(6a) \times h$

$12a^2 = 3ah$

$\frac{12a^2}{3a} = h$

$4a = h$

Step 2: We now use the theorem of Pythagoras to find the hypotenuse of the triangle, which is also the side of the square:

$c^2 = a^2 + b^2$

$h^2 = (6a)^2 + (4a)^2$

$h^2 = 36a^2 + 16a^2$

$h^2 = 52a^2$

$h = \sqrt{52}\ \sqrt{a^2} = \sqrt{52}\ a$

Step 3: We now find the area of the square.

Area = l^2

Area =

SKILLS FOR EXAM SUCCESS MATHS

In the following questions, you will be given the area or the perimeter of the given shape and asked to find the missing dimension.
Note: In this section, you will revise the skills of algebra once more.

3. a) The area of the given triangle is 47·25 cm². The base is 13·5 cm.
        Find the value of h.

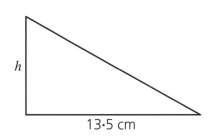

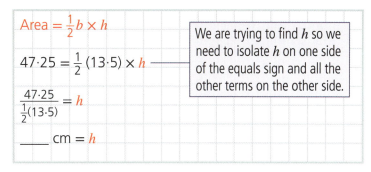

Area = $\frac{1}{2} b \times h$

$47\cdot25 = \frac{1}{2}(13\cdot5) \times h$

$\dfrac{47\cdot25}{\frac{1}{2}(13\cdot5)} = h$

_____ cm = h

We are trying to find h so we need to isolate h on one side of the equals sign and all the other terms on the other side.

b) The circle has an area of 36π cm². Calculate the radius of the circle and hence find the circumference of the circle, correct to 2 decimal places.

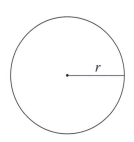

Area = πr^2 Circumference = $2\pi r$
$36\pi = \pi r^2$

c) The perimeter of a rectangle is 150 mm.
    If length : width = 7 : 8, find the length and
    the width of the rectangle.

    Hence find its area.

    Step 1: Ratios (Remember from Chapter 1)
    7 : 8
    $7 + 8 = 15$, therefore the length is $\frac{7}{15}$ and the width is $\frac{8}{15}$.

    Step 2:
    Length: $\frac{7}{15}$ of 150 = _____ mm

    Width: $\frac{8}{15}$ of 150 = _____ mm

    Step 3:
    Area = lw

CHAPTER 12: APPLIED MEASURE

4. Below is a diagram of an isosceles triangle *ABC*. *AD* is the perpendicular bisector of *BC*. |*BD*| = 24 cm. |∠*ABC*| = 40°.

   Calculate:

   a) the perpendicular height of the triangle correct to the nearest whole number

   b) the area of the triangle.

5. The floor of a room is 10 m long and 800 cm wide. The floor is to be covered with tiles, with each tile a square of length 25 cm.

   Applied Arithmetic

   a) How many tiles are needed to cover the floor?

   Step 1: Find the total area of the floor.
   Remember to have all of your dimensions in the same unit!

   Step 2: Find the area of each tile.

   Step 3: Find how many tiles are required to cover the floor.

   b) Each tile costs €5·75 plus VAT at 23%. Calculate the total cost of the tiles, to the nearest euro.

   Find the total cost of the tiles, then add on 23% VAT.

182 SKILLS FOR EXAM SUCCESS MATHS

Section B: Volume, Surface Area and Nets of Rectangular Solids and Cubes

- Volume is the amount of space an object takes up. It is a 3D property.

Rectangular solid (cuboid)

$V = lbh$
$SA = 2lb + 2lh + 2bh$

Cube

$V = l^3$
$SA = 6l^2$

> You need to remember the formulae for volume and area. They are **not** in the *Formulae and Tables* booklet.

- Total surface area is the area of each of the faces of a shape added together.

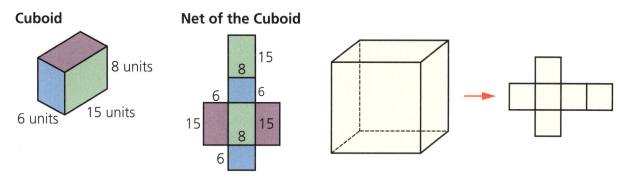

- When calculating volume = units³.
- When calculating area = units².
- 'Net' is the term used to describe what a 3D shape would look like if it was opened out and laid flat.

6. A rectangular block of copper has a height of 4 cm, a length of 6 cm and a width of x cm. The surface area of the block is 228 cm².

   a) Calculate the volume of the rectangular block.

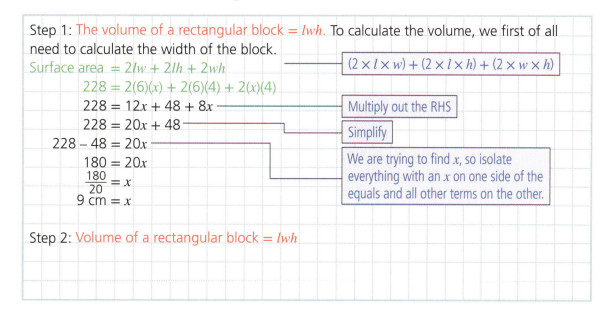

CHAPTER 12: APPLIED MEASURE

b) Calculate how many atoms are in the block of copper if 1 cm³ of copper contains 8·43137 × 10²² atoms. Give your answer correct to 3 significant figures.

Section C: Prisms

- A prism is a 3D shape which has a constant cross-section (a shape that is yielded from a solid when cut by a plane). Both ends have the same 2D shape.

- Cubes and cuboids are also classified as prisms.

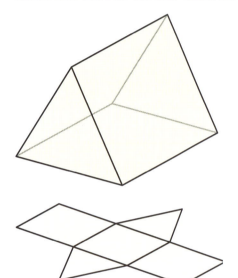

To calculate the surface area of a prism you have to:
1. Identify the shapes that make up each of the sides of the prism.
2. Calculate the area of each individual side.
3. Add all of the areas together.

To calculate the volume of a prism we use the formula:

Cross-sectional area × Length

You need to remember this formula. It is **not** in the *Formulae and Tables* booklet.

In the diagram above, the cross-sectional area is calculated by working out the area of the triangular face.

The length is the base of the parallelogram.

184 SKILLS FOR EXAM SUCCESS MATHS

7. a) Calculate the surface area of the prism.

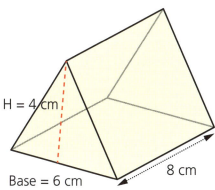

H = 4 cm
Base = 6 cm
8 cm

Triangle: $b = 6$, $h = 4$
Parallelogram: $b = 8$, $h = ?$
Rectangle: $l = 8$, $w = 6$
Before we calculate the area of the parallelogram, we must first calculate the height. The height of the parallelogram is also the hypotenuse of the right-angled triangle. $h^2 = a^2 + b^2$

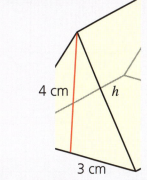

4 cm   h
3 cm

Step 1: Identify the shapes that make up the prism:
2 triangles + 2 parallelograms + 1 rectangle.

Step 2: Find the area of each individual face of the prism and add them altogether.

$$2(\tfrac{1}{2}bh) + 2(bh) + lw$$

b) Calculate the volume of the prism.

Cross-sectional area × Length

Section D: Cylinders, Spheres and Hemispheres

Cylinder
Volume: $V = \pi r^2 h$
Curved surface area: CSA $= 2\pi rh$
Total surface area: TSA $= 2\pi rh + 2\pi r^2$

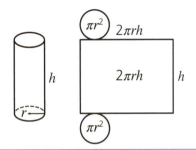

Sphere:
Volume: $V = \frac{4}{3}\pi r^3$
Curved surface area: CSA $= 4\pi r^2$

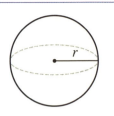

Hemisphere:
Volume: $V = \frac{2}{3}\pi r^3$
Curved surface area: CSA $= 2\pi r^2$
Total surface area: TSA $= 2\pi r^2 + \pi r^2 = 3\pi r^2$

- The formulae for volume and curved surface area of a cylinder and a sphere are in the *Formulae and Tables* booklet.

- The formulae for hemisphere and total surface area of a cylinder are not in the *Formulae and Tables* booklet.

8. A cylinder has a diameter of 15 cm and a height of 20 cm. Calculate the volume, the curved surface area and the total surface area. Give your answers in terms of π.

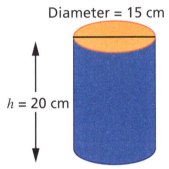

a) the volume

> Volume $= \pi r^2 h$
> Volume $= \pi (7 \cdot 5)^2 (20)$
> Volume $= 1125\pi$ cm³

In the question we are given the diameter. The formula requires the radius. Keep an eye out for this in the exam! Radius $= \frac{1}{2}(15) = 7\cdot 5$ cm.
Because units are not mentioned in the question, you **must** have them in your answer or marks will be lost.

b) the curved surface area

> Curved surface area $= 2\pi rh$. The curved surface area of a cylinder is represented by the blue region in the diagram below. It is the area of the curved part of the cylinder.

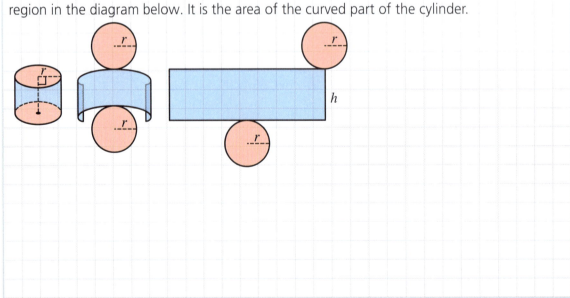

SKILLS FOR EXAM SUCCESS MATHS

c) the total surface area

> Total surface area = $2\pi rh + 2\pi r^2$. The total surface area is the area of the curved portion of the cylinder ($2\pi rh$) **and** the area of the circular top **and** the circular bottom ($2\pi r^2$).

9. The volume of a sphere is 268·08cm³.

   a) Calculate the radius of the sphere, to the nearest cm.

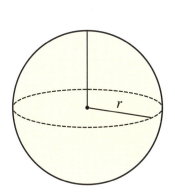

   Volume = $\frac{4}{3}\pi r^3$

   $268·08 = \frac{4}{3}\pi r^3$

   > We are trying to find the radius so we need to isolate r on one side of the equals and all other terms on the other side.

   $\dfrac{268·08}{\frac{4}{3}\pi} = r^3$

   $\sqrt[3]{\dfrac{268·08}{\frac{4}{3}\pi}} = r$

   $3·99999 = r$

   $4 \text{ cm} = r$

   b) Calculate the curved surface area of the sphere, correct to 2 decimal places.

   > Curved surface area = $4\pi r^2$

10. An open cylindrical jar is to be upcycled to make a vase. The radius of the jar is 4·5 cm and it has a height of 25 cm. Calculate the surface area of the jar that has to be painted, correct to the nearest cm².

    > A key point in the question here is that the jar is **open**. This means that the circular bottom of the jar and the curved surface of the jar are to be painted.
    > Formula: $2\pi rh + \pi r^2$

CHAPTER 12: APPLIED MEASURE

11. A cylindrical saucepan, with a radius of 25 cm and height of 40 cm, contains vegetable soup. The chef has to cater for a wedding party of 290 guests. A ladle in the shape of a hemisphere with a radius of 5 cm is used to serve the soup into bowls. Each bowl is to contain one ladle full of soup.

    Will the chef have enough soup to cater for the wedding party?

    Step 1: Find the volume of soup in the cylindrical saucepan.
    Volume = $\pi r^2 h$

    Step 2: Find the volume of soup in the hemispherical ladle.
    Volume = $\frac{2}{3} \pi r^3$

    Step 3: Find the number of ladles of soup contained in the saucepan.

    Step 4: Conclusion. Will the chef have enough soup to feed the wedding party?

    It is important to include your conclusion in your answer so that the examiner can see that you understand what your answer means in the context of the question.

Section E: Compound Volumes

12. A birdcage is in the shape of a hemisphere mounted on a cylinder.
The total height of the cage is 150 cm and it has a radius of 0·5 m.

a) Calculate the total volume of the cage in terms of π, in m³.

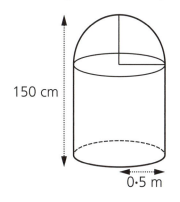

The total height of the cage is 150 cm.
We convert this to metres = 1·5 m.
To find the height of the cylinder,
we must subtract the radius of the
hemisphere from the total height of
the cage.
1·5 m − 0·5 m = 1 m

Step 1: Find the volume of the hemisphere.
$\frac{2}{3}\pi r^3$

Step 2: Find the volume of the cylinder.
$\pi r^2 h$

Step 3: Total volume of the cage:

b) The owner wanted to cover the cage with a blackout cover at night.
Calculate the surface area of the cage that needs to be covered, correct to 2 decimal places.

Curved surface area of the hemisphere +
Curved surface area of the cylinder

CHAPTER 12: APPLIED MEASURE

Section F: Equal Volumes

13. A cylindrical wax candle of radius 6 cm is melted down to form a spherical wax candle of radius 7 cm, with no wax wasted.

Calculate the height of the cylindrical candle, correct to 2 significant figures.

Volume of wax in the cylinder = Volume of wax in the sphere
$$\pi r^2 h = \frac{4}{3}\pi r^3$$

Step 1: Sub the values you know into the formulae.
Step 2: Isolate h on one side of the equals and all other terms to the other.

14. a) A sphere has a radius of 3 cm. Calculate the volume of the sphere in terms of π.

$$\frac{4}{3}\pi r^3$$

b) 10 of these spheres are completely submerged into a cylinder containing water. The cylinder has a radius of 9 cm. By how much will the level of the water rise when the spheres are placed in the cylinder? Give your answer to 2 significant figures.

Step 1: Calculate the volume of 10 such spheres.
$$10\left(\frac{4}{3}\pi r^3\right)$$

Step 2: The volume of the spheres = the volume by which the water rises in the cylinder.
$$10\left(\frac{4}{3}\pi r^3\right) = \pi r^2 h$$

Substitute the values you know into the equation.
You are trying to calculate the height by which the water rose so we need to isolate h on one side of the equals sign and all other terms on the other side.

SKILLS FOR EXAM SUCCESS MATHS

Exam-Style Questions

Exam Hints

1. Attempt every part of every question. You cannot get marks for a blank space!
2. Highlight key words in the question. Make sure you understand what you are being asked to do.
3. Write the relevant formula. The correct formula with some relevant substitution will get you marks.
4. If you are asked to give your answer in a particular format and are unsure of how to do this, do as much as you can.
5. Marks are not awarded just for the correct answer, they are also available for work completed along the way.
6. If units are not mentioned in the question, marks will be awarded for having the correct units in your answer.

Question 1

Linked Question – Applied Measure, Trigonometry, Geometry, Algebra, Applied Arithmetic

A landscaper is designing a garden, *ABCD*, which will consist of grass, a circular patio, two identical right-angled triangular flowerbeds and a water feature in the shape of a sector. His plans are outlined in the diagram below.

a) Calculate the area of the grass in the garden, correct to the nearest m².

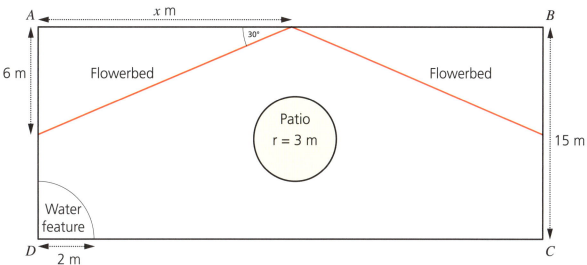

CHAPTER 12: APPLIED MEASURE

In order to calculate the area of the grass in the garden we have to complete the following:
Area of the rectangle ABCD − [2(Area of the triangle) + Area of the circle + Area of the sector]

Step 1: The first calculation we have to complete is to find the value of x, the base of the triangle. We need this value for two reasons: firstly, to find the area of the triangle and secondly, to find the length of the rectangle.

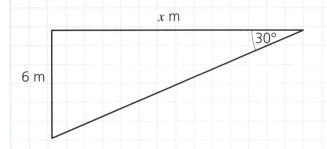

To find x, we use trigonometric functions:
$\tan \theta = \dfrac{\text{opposite}}{\text{adjacent}}$

Step 2: Find the area of the two triangles.
$2(\tfrac{1}{2}bh)$

Step 3: Area of the rectangle
Area: $\tfrac{1}{2}lw$

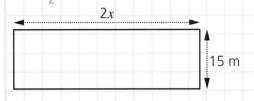

The length of the rectangle is twice the value of x found in Step 1.
Length = $2x$ =
Area of rectangle = lw

Step 4: Area of the patio

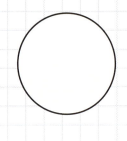

Area of a circle: πr^2

You are not told what value of π to use. In this situation, use the π button on your calculator. The solutions in the marking scheme are usually calculated using the calculator value for π.

Step 5: Area of the water feature, calculated using the area of a sector formula

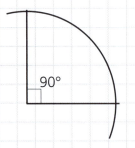

Remember – all angles in a rectangle are 90°.
Area of a sector: $\dfrac{\theta}{360} 2\pi r$

Step 6: Area of the rectangle ABCD − [2(Area of the triangle) + Area of the circle + Area of the sector]

b) The landscaper earns €20 000 for the job.

He has a standard rate cut-off point (SRCOP) of €14 500. The standard rate of tax is 22% and the higher rate of tax is 42%. He has tax credits of €1450.

He also has to pay insurance of €200.

Calculate his net income for the job.

Gross tax = Standard rate on all income up to the SRCOP + Higher rate on all income above the SRCOP
Tax payable = Gross Tax − Tax Credits
Net income = Gross income − Tax payable − Other expenses

Question 2 – 2021

Linked Question – Applied Measure, Numbers

a) A can in the shape of a cylinder has a radius of 3·6 cm and a height of 10 cm.
Work out the volume of the can. Give your answer in cm³, correct to 2 decimal places.

24 of these cans are to be packed into a closed rectangular box.

b) The cans will be arranged inside the box as follows.

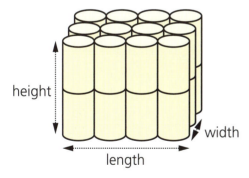

i. Write down the height, the length, and the width of the smallest rectangular box that will be needed for these 24 cans. One is already done for you.

Height = _____   Length = _28·8 cm_   Width = _____

ii. Work out the volume of this box.

CHAPTER 12: APPLIED MEASURE **193**

c) Work out the percentage of the volume of this box that is taken up by the 24 cans.
Give your answer correct to 1 decimal place.

There are a number of different ways of arranging the 24 cans so that they can be packed into a rectangular box. The dimensions of the box may be different for different arrangements.

d) Find the dimensions of the rectangular box for a different arrangement of the 24 cans.
Show your working out.

Height = _____   Length = _____   Width = _____

Question 3

Linked Question – Applied Measure, Geometry, Trigonometry

A packet of sweets is in the shape of a closed triangular-based prism.

It has a height of 8 cm and a triangular base with sides of length 4 cm, 4 cm and 6 cm.

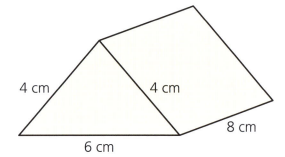

a) Construct an accurate net of the prism.

Show all of your construction lines clearly.

Remember!
The net of a shape is a term used to describe what a 3D shape would look like if it was opened out and laid flat.

194 SKILLS FOR EXAM SUCCESS MATHS

b) Construct a triangle ABC, the front face of the prism, where |AB| = 6 cm and |AC| = |BC| = 4 cm.

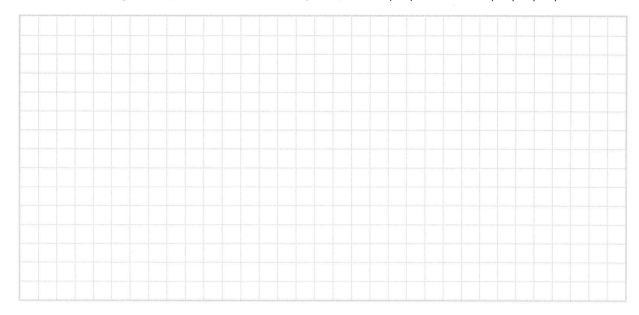

c) Construct the perpendicular bisector of AB. Show all construction lines clearly.

d) Calculate the perpendicular distance from the point C to the side AB.

Hint: $c^2 = a^2 + b^2$

e) Find the total surface area of the prism. Give your answer correct to 2 significant figures.

Show all your workings in the space provided.

f) Calculate the volume of the prism. Give your answer to the nearest whole number.

Summary review:

- Units in answers. If the question asks you to give your answer in m³, for example, no additional marks are given for including m³ in your answer. If there is no reference to units in the question, you will be deducted marks for not correctly stating the units in your answer.

- Make sure your answers are in the correct format, e.g. in terms of π or to the correct number of decimal places.

- I would advise putting the value you are substituting into a formula in brackets, as this will help to reduce any mistakes with signs.

- Many of the formulae you require are on pages 8, 9 and 10 of the *Formulae and Tables* booklet.

- The following formulae are not in the booklet and will need to be remembered.

    - **Rectangle:** Area: lw; Perimeter: $2l + 2w = 2(l + w)$

    - **Square:** Area: l^2; Perimeter: $4l$

    - **Cylinder:** Volume: $\pi r^2 h$; Curved surface area: $2\pi rh$; Total surface area: $2\pi rh + 2\pi r^2$

    - Hemisphere: Volume: $\frac{2}{3}\pi r^3$; Curved surface area: $2\pi r^2$; Total surface area: $2\pi r^2 + \pi r^2 = 3\pi r^2$

- To calculate the surface area of a prism you have to:

    - identify the shapes that make up each of the sides of the prism

    - calculate the area of each individual side

    - add all of the areas together.

- Formula to calculate the volume of a prism: Cross-sectional area × Length. In this diagram, the cross-sectional area is calculated by finding the area of the triangular face. The length is the base of the parallelogram.

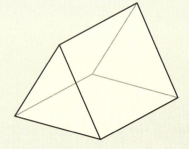

Scan this code to access a video with worked solutions for every question.

Chapter 13: Statistics 1

The topics and skills I will work on are:

Theory:
- Types of data
- Methods of gathering data
- Bias

Bar charts

Line plots

Pie charts
- Drawing a pie chart
- Interpreting information when given the pie chart

Measures of central tendency (average) and spread:
- Mode
- Median
- Mean
- Measures of variability: Range

Section A: Theory

1. Complete the following diagram.

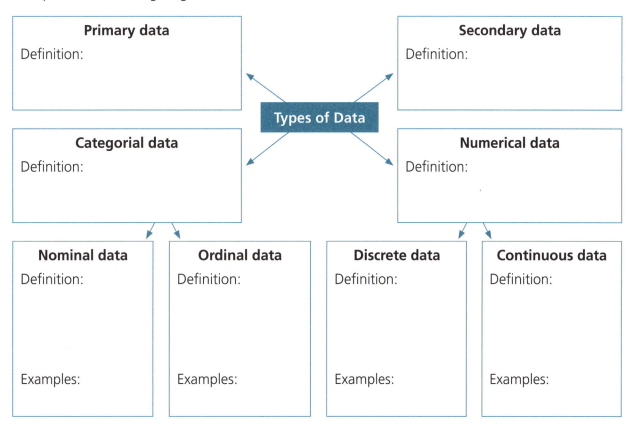

CHAPTER 13: STATISTICS 1   197

2. The following table lists a variety of ways of collecting and gathering data. Outline one other advantage and one other disadvantage of each method of collection.

Method of collection	Advantages	Disadvantages
Questionnaires	1. Questions can be explained 2.	1. Response rate can be low 2.
Direct observation	1. Inexpensive 2.	1. May include human error 2.
Experiments	1. Very accurate data if method followed correctly 2.	1. May include human error 2.
Interviews	1. Questions can be explained 2.	1. Interviewers might influence answers 2.
Databases	1. Inexpensive 2.	1. Information may be out of date 2.

3. Outline three guidelines that should be followed when writing questions for a survey/questionnaire.

   a)

   b)

   c)

4. Give three ways that you can make sure bias does not exist in your survey.

   Bias is an inbuilt error that causes the data to be inaccurate. For example, if you are trying to see how many Manchester United supporters there are in an area but you only ask people who are wearing Manchester United jerseys.

   a)

   b)

   c)

198 SKILLS FOR EXAM SUCCESS MATHS

Section B: Bar Charts

- Bar charts are suitable for categorical **and** numerical discrete data.

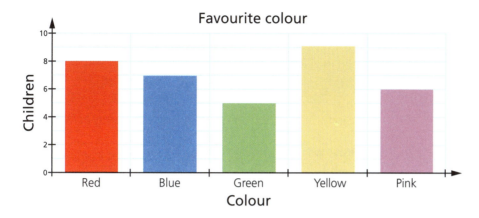

5. A survey was carried out on a group of six-year-old girls. They were asked their favourite after-school activity. The results were as follows:

Activity	Swimming	Dance and Drama	Football	Gymnastics	Piano	Hurling
Number of girls	2	7	5	8	2	6

a) Represent this information on a bar chart.

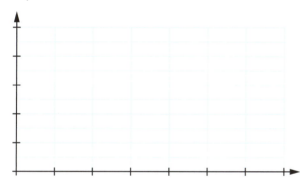

Remember!

When drawing a bar chart you need to:
1. Use a pencil.
2. Label the axes. Here the y-axis will be labelled 'Number of girls' and the x-axis will be labelled 'Activity'.
3. Each bar in the bar chart must be the same width and be clearly labelled.
4. The **space between the bars** must be equal.

b) What type of data is this an example of?
   _____

   Give a reason for your answer.

c) The surveyor stated that this data was an accurate representation of the activities preferred by all children aged 5–9.

   Is this statement correct? Give a reason for your answer.

Exam Hint

This is a two-part question. There are marks available for both parts in the exam.

CHAPTER 13: STATISTICS 1   199

d) A student is picked at random from the group. Find the probability that the student had hurling as her favourite activity.

Give your answer as a fraction in its simplest form.

e) What percentage of the group had swimming or gymnastics as their favourite activity?

Give your answer correct to the nearest whole number.

$$\% = \frac{\text{Swimming + gymnastics}}{\text{Total number of students}} \times 100\%$$

Section C: Line Plots

- Line plots are suitable for categorical **and** numerical discrete data.

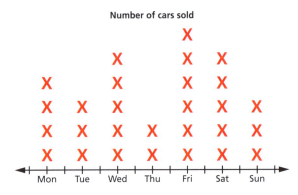

6. A survey was carried out on a group of 25 people asking them their favourite colour. The results were as follows:

Red	Pink	Orange	Pink	Red
Blue	Blue	Red	Purple	Pink
Pink	Red	Pink	Orange	Red
Purple	Pink	Pink	Purple	Blue
Orange	Orange	Blue	Red	Pink

a) Represent this data with a frequency table. The data for 'Red' has been filled in for you.

Frequency tables are suitable for categorical and discrete numerical data.

Colour	Red	Blue	Pink	Purple	Orange
Tally	|||||				
Frequency	6				

Remember:
||||| represents 5 items.

SKILLS FOR EXAM SUCCESS MATHS

b) Represent this data with a line plot.

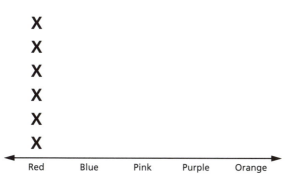

Remember!
When drawing any kind of graph or chart, always use a pencil.
Each X will represent one vote in the survey.
Make sure the Xs are evenly spaced.

c) What type of data is this an example of? _____

Give a reason for your answer.

d) Design a fair question that would generate this type of data.

Section D: Pie Charts

- Pie charts are suitable for categorical **and** numerical discrete data.

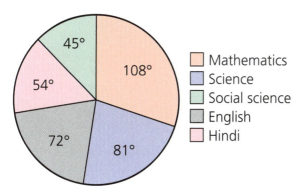

7. 72 fifth-year students were asked which of the following subjects they would select as their number one choice to study for their Leaving Cert.

The results were as follows:

Subject	Biology	Chemistry	Physics	Applied Maths	Agricultural Science
Number of students	20	16	18	10	8

Represent this information on a pie chart.

CHAPTER 13: STATISTICS 1 **201**

Step 1: We need to calculate the number of degrees of the pie chart that are dedicated to each subject.

Biology: 20 out of 72 students surveyed want to study Biology.

There are 360° in a pie chart.

So $\frac{20}{72} \times 360 = 100°$ of the pie chart will represent Biology.

Chemistry: 16 out of 72 students want to study Chemistry.

$\frac{16}{72} \times 360 = 80°$ of the pie chart will represent Chemistry.

Physics: 18 out of 72 students want to study Physics.

So _____ of the pie chart will represent Physics.

Applied Maths: 10 out of 72 students want to study Applied Maths.

So _____ of the pie chart will represent Applied Maths.

Agricultural Science: 8 out of 72 students want to study Agricultural Science.

So _____ of the pie chart will represent Agricultural Science.

Exam Hint

In the exam, there are marks available for finding the number of degrees of the pie chart dedicated to each sector **and** for drawing the pie chart.

Chart title

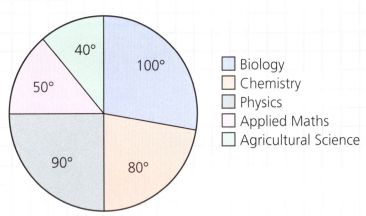

When drawing the pie chart, a tolerance of ± 2° is allowed for each sector.
Use a protractor placed on the start line to measure each of your calculated angles around the circle. Each new line drawn becomes the start line for the next angle.
You need to label each sector with the number of degrees and the data it represents.

8. 36 Leaving Certificate students were asked where they were going on their Leaving Certificate holiday.

The results are represented in the table below:

Destination	Marbella	Santa Ponsa	Ibiza	Magaluf	Ayia Napa
Number of students	4	9	6	12	5

202 SKILLS FOR EXAM SUCCESS MATHS

a) Represent this information on a pie chart.

Step 1: Calculate the number of degrees of the pie chart for each holiday destination.
Marbella:

Santa Ponsa:

Ibiza:

Magaluf:

Ayia Napa:

Step 2: Draw the pie chart.

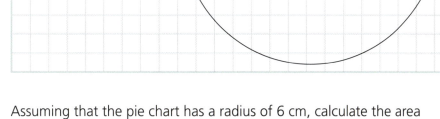

Applied Measure

b) Assuming that the pie chart has a radius of 6 cm, calculate the area of the sector of the pie chart represented by:

Remember: area = units²

i. students who went to Magaluf

ii. students who went to Santa Ponsa.

Give your answers correct to the nearest whole number.

Area of a sector: $\frac{\theta}{360}\pi r^2$

CHAPTER 13: STATISTICS 1 | 203

9. The pie chart below illustrates the grades obtained by a group of students who completed an entrance assessment for their local secondary school.

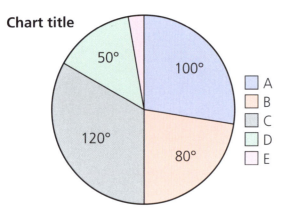

a) What is the angle representing Grade E?

b) 50 students were awarded Grade A. Calculate the number of students in each of the remaining categories.

Step 1:
Grade A: 100° = 50 students

$$1° = \frac{50}{100}$$

$$1° = 0.5 \text{ students}$$

Step 2:
Grade B: 80° = 80 × 0.5 = 40 students
Grade C: 120° =
Grade D: 50° =
Grade E: 10° =

c) How many students sat the exam? _____

d) What type of data is this? _____

204 SKILLS FOR EXAM SUCCESS MATHS

Section E: Measures of Central Tendency (Average) and Spread

The mode

This is the value that occurs the most often.

Example: 1, 2, 3, 3, 4, 4, 4, 5

In this set of data, the mode is 4 as it occurs three times. It is the number that occurs the most often.

> Advantage: Not affected by outliers.
> Disadvantage: There can be more than one answer or there may be no mode at all.

The median

This is the middle value when the data is in **ascending order**. It divides the data exactly in half.

50% of the data is below the median, 50% of the data is above the median.

Example: 3, 4, 5, 6, 7

Here the median is 5.

Example: 1, 2, 3, 3, 4, 4, 4, 5

Here there are two middle values, 3 and 4.

To calculate the median, we add 3 and 4 together and divide our answer by 2.

$\frac{3+4}{2} = 3.5$

> Advantage: Not affected by outliers.
> Disadvantage: Not as well understood as the mean.

> $\frac{1}{2}(n+1)$ can also be used to find the median.

The mean

This is the sum of all the values divided by the number of values.

Example: 1, 2, 3, 3, 4, 4, 4, 5

$\frac{1+2+3+3+4+4+4+5}{8} = \frac{26}{8} = 3.25$

> Advantage: It uses all data points and does not exclude any values.
> Disadvantage: Affected by outliers.

The range

This is the Highest value − Lowest value.

Example: 1, 2, 3, 3, 4, 4, 4, 5

$5 - 1 = 4$

CHAPTER 13: STATISTICS 1

> **Worked example**
>
> Find the mean, mode, median and range of the following set of data:
>
> 25, 16, 14, 19, 13, 15, 17, 14, 18, 14
>
> **Mean:**
>
> $$\frac{25 + 16 + 14 + 19 + 13 + 15 + 17 + 14 + 18 + 14}{10}$$
>
> $$= \frac{165}{10} = 16{\cdot}5$$
>
> **Mode:**
>
> The mode is the piece of data that occurs the most often.
>
> In this set of data the mode is 14.
>
> **Median:**
>
> Before we start calculating the median, the data must first be arranged in ascending order.
>
> 13, 14, 14, 14, 15, 16, 17, 18, 19, 25
>
> To find the median here, we must add 15 and 16 and divide our answer by 2.
>
> $$\frac{15 + 16}{2} = 15{\cdot}5$$
>
> **Range:**
>
> Largest value – Smallest value
>
> 25 – 13 = 12

10. a) What is an outlier? Identify the outlier in the set of data in the example above.

    b) Which measure of central tendency does the outlier have an effect on?

    c) Will outliers have an effect on the range? Give a reason for your answer.

11. You are asked to analyse the following set of data. Answer the questions that follow.

    21, 10, 25, 23, 24, 27, 25, 26

    a) Calculate the mean of the data.

b) Calculate the mode of the data.

c) Calculate the median of the data.

d) In your opinion, which one of mean, mode or median gives the most accurate representation of the measure of central tendency?

   Explain your answer.

e) Calculate the range of the data.

f) What does the range tell us about the data?

12. The results in an English exam were given as 98, 95, x, 88, 92, 86, 96 and 80.

    The mean result was 89·875. Calculate the value of x.

    Hint: Form an algebraic equation with the data.

CHAPTER 13: STATISTICS 1 **207**

Exam-Style Questions

Exam Hints

1. Attempt every part of every question. You cannot get marks for a blank space!
2. Highlight key words in the question. Make sure you understand what you are being asked to do.
3. If you are asked to give your answer in a particular format and are unsure of how to do this, do as much as you can.
4. Marks are not just given for the correct answer, they are also available for work completed along the way.

Question 1

Linked Question – Statistics, Algebra, Real Numbers

a) Write five integers on the lines below so that the set of numbers has:

– a mode of 2

– a median of 2

– a mean of 4

– a range of 15.

_____ _____ _____ _____ _____

Remember!
Integers are positive and negative whole numbers.

b) Another natural number is added to the list.
The mean is now increased to 5.

Using algebra, calculate the value of this extra number.

Remember!
Natural numbers are positive whole numbers.

c) Write any three natural numbers on the lines below so that the set of numbers has:

– a mode of 3

– a median of 3

– a mean of 5.

_____ _____ _____

208 | SKILLS FOR EXAM SUCCESS MATHS

Question 2 – 2022

Linked Question – Probability, Statistics, Real Numbers

When Maeve's team play a match, they can win (W), lose (L) or draw (D).

a) Fill in the table below to show the 9 possible outcomes when Maeve's team play two matches. One is already done. **W D** means they win Match 1 and draw Match 2.

		Match 2		
		W	D	L
Match 1	W		W D	
	D			
	L			

b) Maeve thinks that each outcome in the table is equally likely.

Based on this, find the probability that, when Maeve's team plays two matches, they **win at least one** match. Give your answer as a fraction.

c) Maeve's team play five matches in a competition.

Work out the total number of different possible outcomes for her team for these matches. For example, one possible outcome would be **W W L D W**.

Maeve's team plays 11 matches in a league. The table below shows the number of goals that Maeve's team score in each of these 11 matches.

3	1	1	0	2	7	1	0	2	1	3

d) Work out the mean number of goals that Maeve's team score per match.

Give your answer correct to 1 decimal place.

e) Complete the pie chart below, to summarise the data above, showing the proportion of their games in which Maeve's team scored 0 goals, 1 goal, and so on.

Label each sector **and** the size of the angle clearly. Show all workings out and construction lines.

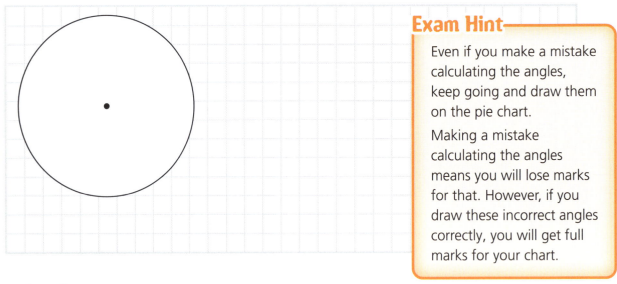

Exam Hint

Even if you make a mistake calculating the angles, keep going and draw them on the pie chart.

Making a mistake calculating the angles means you will lose marks for that. However, if you draw these incorrect angles correctly, you will get full marks for your chart.

Question 3

Linked Question – Statistics, Arithmetic, Probability, Real Numbers

A survey was carried out among a group of students. They were asked who their favourite music artist was.

The results were presented on a pie chart.

Chart title

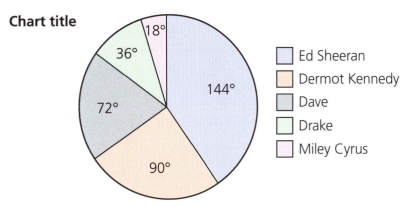

a) Complete the table to show the number of people who voted for each artist.
Show all your workings in the space provided.

Artist	Ed Sheeran	Dermot Kennedy	Dave	Drake	Miley Cyrus
Degrees in the pie chart	144°	90°	72°	36°	18°
Number of votes		25			

Hint:
90° = 25 votes
$1° = \frac{25}{90}$

210　SKILLS FOR EXAM SUCCESS MATHS

b) Represent the number of votes each artist received on a suitable alternative chart.

c) i. What type of data is this? Tick one box only.

   Nominal   Ordinal   Discrete   Continuous
   ☐   ☐   ☐   ☐

   ii. Give a reason for your answer.

d) What percentage of students surveyed voted for Ed Sheeran **or** Dermot Kennedy?

e) Another 50 students are surveyed. If the probability of voting for Drake is 0·1, how many of these 50 students would you expect to **not vote** for Drake?

Summary review:

- The mode: the value that occurs the most often.
- The median: the middle value when the data is in ascending order.
- The mean: the sum of all the values divided by the number of values.
- The range: Highest value − Lowest value.
- Bar charts: used to represent categorical and numerical discrete data.
- Line plots: used to represent categorical and numerical discrete data.
- Pie charts: used to represent categorical and numerical discrete data.

Scan this code to access a video with worked solutions for every question.

CHAPTER 13: STATISTICS 1 — 211

Chapter 14: Statistics 2

The topics and skills I will work on are:

Central tendencies in a grouped frequency distribution: ☐
- **Mean** ☐
- **Mode** ☐
- **Median** ☐

Histograms ☐

Stem and leaf diagrams ☐

Back-to-back stem and leaf diagrams ☐
- **Measures of central tendency in stem and leaf diagrams** ☐
- **Measure of variability: range** ☐

Section A: Central Tendencies in a Grouped Frequency Distribution

- Continuous data is data that can take on any value within a range. It is data that can be measured, e.g. height, weight, temperature, rainfall.

Worked example

A survey was carried out on 60 first-year students.

They were asked how long they had spent doing their homework the previous evening.

The results are represented in the following grouped frequency distribution table.

Time (minutes)	25–30	30–35	35–40	40–45	45–50
Number of students	6	15	20	9	10

a) Calculate the mean of the data.

Remember: The interval 25–30 means from 25 up to but not including the value of 30. 30–35 means from 30 up to but not including 35 and so on.

Step 1: In order to calculate the mean of a grouped frequency distribution, we need to calculate the mid-interval value in each category.

The mid-interval value is the middle value in each individual group. For example, the mid-interval value in the 25–30 group is 27·5. It is the number that lies halfway between 25 and 30.

Time (minutes)	25–30	30–35	35–40	40–45	45–50
Mid-interval value	27·5	32·5	37·5	42·5	47·5
Number of students	6	15	20	9	10

212 SKILLS FOR EXAM SUCCESS MATHS

Step 2: Now that we know the mid-interval values for each group, we can calculate the mean.

We multiply the mid-interval value by the number of students for each category, add all this data together and divide by the number of pupils surveyed.

$$\frac{(27{\cdot}5 \times 6) + (32{\cdot}5 \times 15) + (37{\cdot}5 \times 20) + (42{\cdot}5 \times 9) + (47{\cdot}5 \times 10)}{6 + 15 + 20 + 9 + 10}$$

$$\frac{165 + 487{\cdot}5 + 750 + 382{\cdot}5 + 475}{60}$$

$$\frac{2260}{60} = 37{\cdot}67 \text{ mins}$$

b) Calculate the mode interval of the data.

The mode is the piece of data that occurs most often, it is the interval that contains the most students.

Here the interval that contains the most students (20) is 35–40 minutes.

c) Calculate which interval the median lies in.

60 students were surveyed.

The median will be the average of the 30th and 31st values.

Time (minutes)	25–30	30–35	35–40	40–45	45–50
Number of students	6	15	20	9	10

6 + 15 = 21   21 + 20 = 41

The average of 30 and 31, 30·5, lies in the 35–40 group.

Therefore, the median lies in the 35–40 group.

Exam Hint

I appreciate all of this work can be done on the calculator. However, I would advise that you write out each step as shown here. This way, any error can be pinpointed by the examiner and minimal marks will be lost.

In some instances, the correct answer with no workings can result in only LPC or no marks being awarded.

Section B: Histograms

- Histograms are suitable for plotting continuous numerical data.

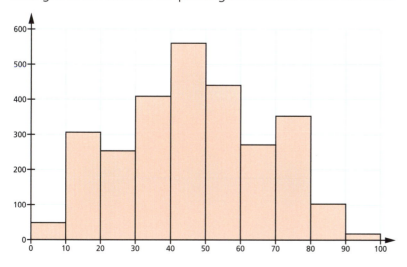

CHAPTER 14: STATISTICS 2

1. The table below shows the time taken for a group of people to complete their driver theory test.

   a) Represent this information on a histogram.

Time (minutes)	30–32	32–34	34–36	36–38	38–40	40–42
Number of people	4	6	3	7	3	1

   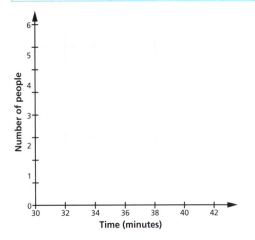

   Remember!
   When drawing a histogram:
   - Label the axes.
   - Always use a pencil and a ruler.
   - Make sure each bar is the same width.
   - In the histogram there is **no space** between the bars. They are joined together. This is because the data being represented is continuous data.

   b) How many people are in the group?
   _____

   c) Using mid-interval values, calculate the mean time it took the group to complete the test.

   d) What is the modal interval? _____

   e) In which interval does the median lie?

   f) What percentage of people completed the test in less than 38 minutes? Give your answer correct to 2 decimal places.

   $$\% \text{ of people} = \frac{\text{People who completed exam in} < 38 \text{ mins}}{\text{Total number of people}} \times 100\%$$

SKILLS FOR EXAM SUCCESS MATHS

g) What is the **maximum** number of people who completed the test in greater than 37 minutes?

h) What is the **minimum** number of people who could have completed the test in greater than 37 minutes?

i) Give one advantage and one disadvantage of using a histogram to represent statistical data.

Advantage:

Disadvantage:

Section C: Stem and Leaf Diagrams

- Stem and leaf diagrams are suitable for representing numerical data, both discrete and continuous.

Key : 2	0 means 20
Stem	Leaf
0	1 4
1	3 6 6 6 7
2	0 2 5
3	6 7 7 7 8
4	0 1 3

Exam Hint

To ensure you get full marks for your stem and leaf diagrams you need to make sure:
- the data is in ascending order
- all data is included
- the diagram has a key.

Marks will be lost if you do not meet these criteria.

2. The ages of a group of workers in a company are listed in the table below.

18	52	62	22	19	55	65	20
21	30	19	34	45	39	43	58
40	60	36	45	34	63	45	25
27	34	44	53	34	19	56	46
51	60	54	26	54	47	22	27

CHAPTER 14: STATISTICS 2

a) Illustrate this information on a stem and leaf diagram.

Step 1: To draw a stem and leaf diagram, the data must be in ascending order.
The first thing that you need to do is order the data.
18, 19, 19, 19
20, 21, 22, 22, 25, 26, 27, 27
30, 34, 34, 34, 34, 36, 39
40, 43, 44, 45, 45, 45, 46, 47
51, 52, 53, 54, 54, 55, 56, 58
60, 60, 62, 63, 65

Remember: Marks will be lost in the exam if the data in the stem and leaf diagram is not in the correct order.

Step 2: We are now ready to put the data into a stem and leaf diagram.

Any number can be used to illustrate the key. I recommend using the first or the last number of the stem and leaf diagram as the key.

Key: 1|8 = 18

b) Calculate the mean of the data in the stem and leaf diagram.

c) Calculate the median of the data.

d) Explain what the median means in the context of this question.

e) Calculate the mode of the data.

216 SKILLS FOR EXAM SUCCESS MATHS

f) Calculate the range of the data.

Section D: Back-to-Back Stem and Leaf Diagrams

- Back-to-back stem and leaf diagrams are a useful way of comparing two sets of data.

3. A class of 15 students sat a maths exam on the same day. A week later they took a second maths exam. Their scores are shown in the tables below.

   a) Draw a back-to-back stem and leaf diagram to display the students' scores.

   Test 1

63	64	65	66	70
70	70	73	75	79
88	89	90	95	95

   Test 2

68	69	70	74	74
75	76	78	81	84
84	86	96	98	98

   Step 1: The data must be in the correct order. This has already been done for you in this example.

   Step 2: Complete the back-to-back stem and leaf diagram. The first row has been completed for you.

Test 1		Test 2
6 5 4 3	6	8 9
	7	
	8	
	9	

   Key:

   The data on the left goes backwards from the stem and leaf diagram.

b) Answer the following questions based on your back-to-back stem and leaf diagram above.

   i. Calculate the mean, the mode and the median of Test 1.

   ii. Calculate the percentage of students who got 75 marks or higher in Test 1.
   Give your answer to the nearest whole number.

   iii. Calculate the mean, mode and median of Test 2.

   iv. Explain what the median means in the context of this question.

   v. Paul says that every student in the class must have done better in Test 2 compared to Test 1. Is he correct?
   Explain your answer.

   Exam Hint
   There are two parts to this question. Remember to answer both.

Exam-Style Questions

Exam Hints

1. Attempt every part of every question. You cannot get marks for a blank space!
2. Highlight key words in the question. Make sure you understand what you are being asked to do.
3. If you are asked to give your answer in a particular format and are unsure of how to do this, do as much as you can.
4. Marks are not just given for the correct answer, they are also available for work completed along the way.

Question 1

Linked Question – Statistics, Algebra, Distance/Time/Speed, Arithmetic, Real Numbers

15 boys and 15 girls went on a school trip to London.

The weight of each student's bag (in kg) is shown in the tables below.

Girls

5·8	6·3	6·9	7·6
7·8	8·0	8·1	8·7
9·1	9·4	9·5	9·6
9·8	9·8	x	

Boys

5·9	6·8	7·4	8·5
8·6	8·7	8·8	9·2
9·4	9·5	9·5	9·7
9·7	10·5	11·1	

a) The mean weight of the girls' bags was 8·5 kg. Find the value of x, the weight of the missing bag.

b) Calculate the mean weight of the boys' bags correct to 1 decimal place.

c) Represent the information on a back-to-back stem and leaf diagram.

Girls		Boys
	5	
	6	
	7	
	8	
	9	
	10	
	11	

Key:

d) Calculate the median weight of the boys' bags and the girls' bags.

Girls' bags: _____

Boys' bags: _____

e) Eoin says: 'In general, the girls' bags are heavier than the boys' bags.'
Based on the above data, is he correct? Give a reason for your answer.

Hint: Refer to the mean and the median weights of the boys' and the girls' bags in your answer.

f) Chloe wants to exchange €200 for pounds sterling at the airport.

The exchange rate on the day is €1 = £0·89.

How much sterling does Chloe receive?

g) The flight time from Cork to London Heathrow is 1 hour and 20 mins. The distance is 688 km.

Calculate the average speed the aeroplane is travelling at, correct to the nearest whole number.

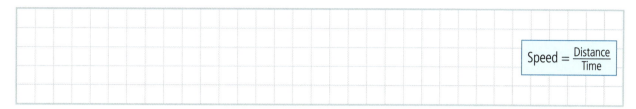

$$\text{Speed} = \frac{\text{Distance}}{\text{Time}}$$

Question 2

Linked Question – Statistics, Co-ordinate Geometry, Geometry, Arithmetic, Real Numbers

Charles De Gaulle Airport and Disneyland Paris are represented on the co-ordinate plane below.

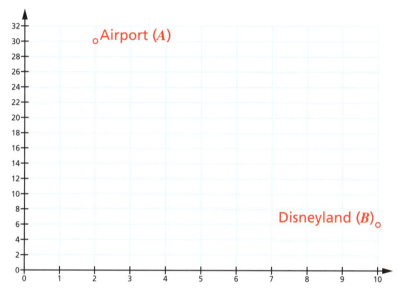

a) Identify the co-ordinates of Point A (the airport) and Point B (Disneyland).

$A = $ _____ $B = $ _____

b) Use this information to calculate the distance between Point A and Point B.

$$d = \sqrt{(x_2 - x_1)^2 + (y_2 - y_1)^2}$$

c) Each unit on the co-ordinate plane represents 1·9 km. Calculate the actual distance from the airport to Disneyland correct to 1 decimal place.

CHAPTER 14: STATISTICS 2   **221**

d) 50 people were asked how long they had spent waiting in line for Thunder Mountain. The results of the survey are given below:

10	45	52	56	43	8	45	23	48	28
21	12	32	15	35	46	49	48	45	58
40	50	45	38	42	58	18	35	49	25
42	34	36	45	22	28	47	55	9	57
5	50	25	40	52	45	55	25	58	35

Complete the grouped frequency distribution below to summarise the data.

Time (mins)	0–10	10–20	20–30	30–40	40–50	50–60
Frequency	3	4	8	7	17	11

e) Represent the information on a histogram.

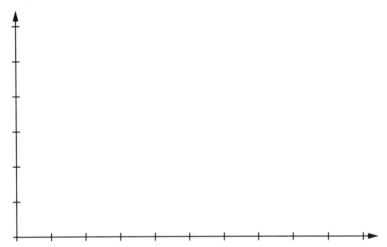

f) Using mid-interval values, calculate the mean waiting time for Thunder Mountain.

g) In what interval does the median lie?

h) A Cinderella costume in the gift shop costs €49·99, inclusive of VAT at 20%.

Calculate the price of the costume before VAT was added, correct to 4 significant figures.

Question 3

Linked Question – Statistics, Probability, Real Numbers

100 students were asked: 'Approximately how long did you spend on social networking sites last week?' The histogram illustrates the answers given by the students.

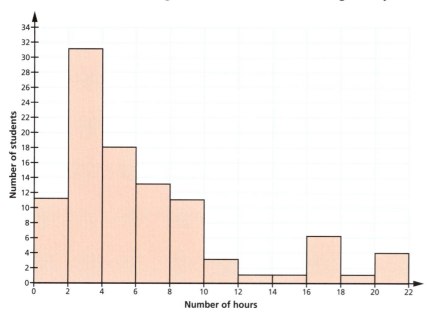

a) Use the data from the histogram to complete the frequency table.

2–4 means 2 hours or more but less than 4 hours, etc.

Number of hours	0–2	2–4	4–6	6–8	8–10	10–12	12–14	14–16	16–18	18–20	20–22
Number of students											

b) What is the modal interval? _____

c) Using mid-interval values, calculate the mean amount of time spent on social networking sites.

CHAPTER 14: STATISTICS 2   **223**

d) A student is picked at random from the group. Find:

  i. the probability that the student picked spent 2–4 hours on social networking sites

  ii. the probability that the student picked spent 16–18 hours on social networking sites

  iii. the percentage of students that spent more than 16 hours on social networking sites during the week.

e) The question was asked to 100 first-year boys the Monday after the midterm break. Give two reasons why the results from the survey might not be entirely accurate of the population in general.

  1.

  2.

Summary review:

- To calculate the mean of a grouped frequency distribution, you need to:
  - calculate the mid-interval value in each category
  - multiply the mid-interval value by the number of values for each category
  - add all this data together and divide by the number of people surveyed.
- You need to remember how to draw a histogram and a stem and leaf diagram.
- You also need to be able to find the mean, mode, median and range of data in a histogram and stem and leaf diagram.

Scan this code to access a video with worked solutions for every question.

224 SKILLS FOR EXAM SUCCESS MATHS

Chapter 15: Probability

The topics and skills I will work on are:

Methods of listing outcomes and fundamental principle of counting
- **Listing**
- **Two-way tables**
- **Tree diagrams**

Probability scale

Relative frequency

Theoretical probability
- **Probability of an event occurring**
- **Probability of an event not occurring**

Expected frequency

Independent events

Sets and probability

Section A: Methods of Listing Outcomes and Fundamental Principle of Counting

- An outcome is a possible result from an experiment or trial.
- We will look at three ways of listing all the outcomes of a trial: listing outcomes, two-way tables and tree diagrams.

Worked example

Kevin is trying to decide which meal option to go for in a restaurant. He has a choice of pasta, pizza or vegetarian for his main course. He can choose ice cream, a fruit plate or chocolate cake for dessert.

How many outcomes are possible?

Give all the possible meals that Kevin could order.

Number of outcomes = 3 options for mains × 3 options for dessert

$$= 3 \times 3$$
$$= 9$$

> **Fundamental principle of counting:**
> If there are m ways of doing one thing and n ways of doing another, then there are $m \times n$ ways of doing both. The number of possible outcomes $= m \times n$.

There are three ways that we can present these results.

1. **Listing**

Pasta and Ice cream	Pasta and Fruit plate	Pasta and Chocolate cake
Pizza and Ice cream	Pizza and Fruit plate	Pizza and Chocolate cake
Vegetarian and Ice cream	Vegetarian and Fruit plate	Vegetarian and Chocolate cake

2. **Two-way table**

	Ice cream	Fruit plate	Chocolate cake
Pasta	Pasta, Ice cream	Pasta, Fruit plate	Pasta, Chocolate cake
Pizza	Pizza, Ice cream	Pizza, Fruit plate	Pizza, Chocolate cake
Vegetarian	Vegetarian, Ice cream	Vegetarian, Fruit plate	Vegetarian, Chocolate cake

3. **Tree diagram**

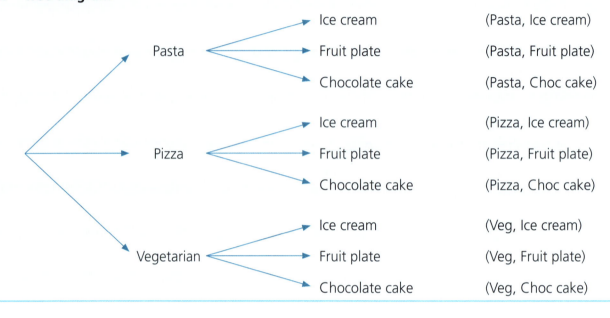

1. A six-sided die is rolled and a coin is flipped.

   a) How many outcomes are possible? _____

   b) Use a tree diagram to list all the possible outcomes.

226 SKILLS FOR EXAM SUCCESS MATHS

2. A code is made up of a vowel (A, E, I, O, U) and a digit (0, 1, 2, 3, 4, 5, 6 ,7, 8, 9).

   Use the fundamental principle of counting to calculate how many outcomes are possible.

   Use a two-way table to list all the possible outcomes.

3. To play a game of Twister, you must spin 2 spinners. Spinner 1 has 4 segments, yellow, blue, green and red. Spinner 2 has segments labelled right hand, right foot, left hand and left foot.

   How many possible outcomes are there? List all the possible outcomes.

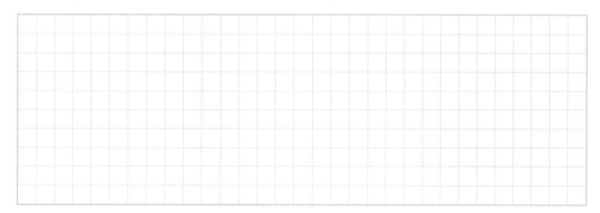

Section B: Probability Scale

4. The following terms can be used to describe the probability that an event happens:

   Impossible   Unlikely   Equally likely   Likely   Certain

   a) For each event in the table below, use one of the terms to describe the probability that it happens.

A.	When a fair coin is tossed, you get a tail.	
B.	If you buy a lottery ticket for next weekend's draw, you will win.	
C.	25 December will be Christmas Day.	
D.	Mexico will win the All Ireland Football Final this year.	
E.	You will pick a red ball from a bag containing 8 red balls and 2 black balls.	

CHAPTER 15: PROBABILITY

b) Write each of the letters A, B, C, D and E into the correct box on the probability scale below to show the probability of each event from part (a). The scale goes from 0 = impossible to 1 = certain.

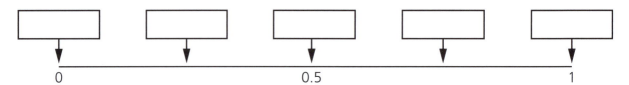

Section C: Relative Frequency

Remember!
You need to remember this formula as it is not in the *Formulae and Tables* booklet.

- To calculate relative frequency, you use the following formula:

Relative frequency = $\dfrac{\text{Number of times the event happens}}{\text{Total number of trials}}$

Worked example
A survey shows that 20 out of 50 households owns a dog. Calculate the relative frequency of a household owning a dog.

Relative frequency = $\dfrac{\text{Number of times the event happens}}{\text{Total number of trials}}$

Relative frequency = $\dfrac{20}{50} = \dfrac{2}{5}$

Always give your answer in its simplest form.

5. a) A soccer team has won 5 out of 8 matches. Find the probability that it will **lose** its next game. Give your answer as a percentage.

b) How could you improve the accuracy of this prediction?

6. A record was kept of the number of days 3 students were late for school over a period of 4 weeks (20 days).

Estimate the relative frequency of each student being late for school. Give your answer as a fraction in its simplest form, as a decimal and as a percentage.

Student	Number of days late	Fraction	Decimal	Percentage
Seán	5			
Megan	8			
Konrad	12			

7. A six-sided die is thrown 30 times.

   a) Calculate the relative frequency of rolling a 1, 2, 3, 4, 5 and 6 based on the data in the table below.

Number on the die	1	2	3	4	5	6
Number of outcomes	5	4	10	5	5	1
Relative frequency						

   b) Is this a fair die? Give a reason for your answer.

   If a die is fair, each number will have an equal chance of occurring.

Section D: Theoretical Probability

- To calculate theoretical probability, you use the following formula:

$$\text{Theoretical probability} = \frac{\text{Number of outcomes of interest}}{\text{Total number of outcomes}}$$

Remember! You need to remember this formula as it is not in the *Formulae and Tables* booklet.

Worked example

A bag contains 10 marbles: 5 red, 3 blue and 2 green.

a) Find the probability of picking:

   i. a red marble from the bag

   $$\text{Probability} = \frac{\text{Number of outcomes of interest}}{\text{Total number of outcomes}}$$

   $$\text{Probability} = \frac{\text{Number of red marbles}}{\text{Total number of marbles}} = \frac{5}{10} = \frac{1}{2}$$

   Always give your answer in its simplest form.

   ii. a blue marble from the bag

   $$\text{Probability} = \frac{\text{Number of blue marbles}}{\text{Total number of marbles}} = \frac{3}{10}$$

   iii. a green marble from the bag

   $$\text{Probability} = \frac{\text{Number of green marbles}}{\text{Total number of marbles}} = \frac{2}{10} = \frac{1}{5}$$

   iv. a marble that is <u>not</u> green.

   The probability that the marble is <u>not</u> green also means what is the probability that the marble <u>is</u> red or blue?

   $1 - \frac{1}{5} = \frac{4}{5}$

b) What is the probability of getting a red marble followed by a green marble, assuming that the marbles are not replaced?

   The probability of picking a red marble from the bag: $\frac{5}{10} = \frac{1}{2}$.

   If the marbles are not replaced, there are 9 marbles left in the bag, 2 of which are green.

   The probability of picking a green marble from the bag is now $\frac{2}{9}$.

   Therefore the probability of picking a red marble followed by a green marble is

   $\frac{1}{2} \times \frac{2}{9} = \frac{2}{18} = \frac{1}{9}$

CHAPTER 15: PROBABILITY

8. A deck of cards is made up of 52 cards.

   There are 4 suits: Hearts, Diamonds, Spades and Clubs.

   There are 13 cards in each suit: 2, 3, 4, 5, 6, 7, 8, 9, 10, Jack, Queen, King, Ace.

   Hearts ♥ :   2, 3, 4, 5, 6, 7, 8, 9, 10, J, Q, K, A

   Diamonds ♦ : 2, 3, 4, 5, 6, 7, 8, 9, 10, J, Q, K, A

   Clubs ♣ :   2, 3, 4, 5, 6, 7, 8, 9, 10, J, Q, K, A

   Spades ♠ :   2, 3, 4, 5, 6, 7, 8, 9, 10, J, Q, K, A

   A card is picked at random from a deck of cards. Calculate the probability of the following:

   a) Picking a red card from the deck

   b) Picking a picture card from the deck

   c) Picking a prime number from the deck

   A prime number is a number that only 1 and the number itself can divide into.

   d) Picking a card whose number is divisible by 3 from the deck

   e) Picking a spade from the deck

   f) Picking a card that is **not a spade** from the deck

9. The probability that it will rain tomorrow is 0·8.

   What is the probability that it will **not** rain?

   There is a decimal in the question so give your answer as a decimal.

SKILLS FOR EXAM SUCCESS MATHS

10. The probability that Layla will miss a penalty is 30%.

    What is the probability that she will score the penalty?

    > There is a percentage in the question so give your answer as a percentage.

11. There are 24 students in a class. The table shows the number of boys and the number of girls, the number who are left-handed and the number who are right-handed.

	Righted-handed	Left-handed
Girls	8	3
Boys	9	4

    A student is picked at random from the class. Find the probability that:

    a) A girl is picked

    b) A boy who is left-handed is picked

    c) A person who is **not right-handed** is picked

    d) A boy who is **not left-handed** is picked.

Section E: Expected Frequency

Remember!
You need to remember this formula as it is not in the *Formulae and Tables* booklet.

- Expected frequency = Number of trials × Relative frequency or Probability

Worked example

A fair-sided die is rolled 500 times. How many times would you expect to roll a prime number?

There are three prime numbers on a standard die: 2, 3 and 5. Relative frequency of rolling a prime number = $\frac{3}{6} = \frac{1}{2}$

Expected frequency = Number of trials × Relative frequency or Probability

$= 500 \times \frac{1}{2}$

$= 250$

CHAPTER 15: PROBABILITY

12. A card is selected at random from a deck of cards (52 cards). This is done 780 times.

    How many times would you expect to get:

    a) A red card _____

    b) A diamond _____

    c) A picture card _____

    d) The king of spades _____

    e) An ace _____

Section F: Independent Events

- Independent events are events where the probability of one event occurring has no effect on the probability of another event occurring.

13. There are 2 spinners. Spinner A has the numbers 2, 4, 6 and 8. Spinner B has the numbers 3, 5, 7 and 9. The spinners are spun and the numbers are added together. Complete the table to show all the possible scores.

	2	4	6	8
3	5			
5		9		
7			13	
9				17

a) How many outcomes are possible? _____

b) Find the probability that the score on the spinners will add up to:

   i. The number 11 _____

   ii. An even number _____

   iii. A multiple of 3 _____

   iv. A number less than 10 _____

   v. A number that is equal to or greater than 15 _____

14. On his way to school, Mr O'Brien passes through 3 sets of traffic lights. There are 2 possible outcomes for each set of lights, Red or Green.

Use a tree diagram to list all the possible outcomes.

List of outcomes

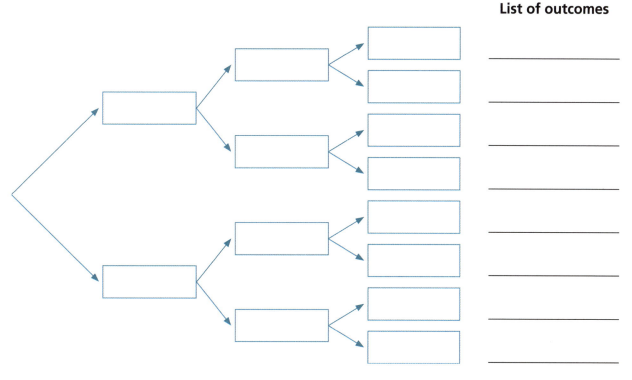

Calculate the probability of:

a) Getting 3 green lights _____

b) Getting just 1 red light _____

c) Getting at least 2 green lights _____

d) Getting only 2 green lights _____

> At least 2 green lights means 2 or 3 green lights.
> Only 2 green lights means only 2 of the lights are green.

Section G: Sets and Probability

15. There are 30 students in a class. 12 study French, 16 study Spanish and 7 study neither.

    a) Fill in the Venn diagram below.

    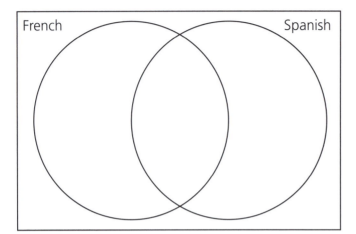

    Remember!
    When filling in a Venn Diagram, first fill in the region that represents A ∩ B, those who study both subjects. Then fill in those who study just French, then those who study just Spanish. Finally, add those who study neither.

    b) One student is picked at random from the class. Find the probability that the student picked:

    i.   Studies both subjects _____

    ii. Studies neither subject _____

    iii. Studies French only _____

    iv. Studies Spanish only _____

    v.   Studies at least one language _____

    vi. Studies one language only _____

16. A group of 100 students were surveyed and asked if they had drunk tea (T), coffee (C) or a soft drink (D) at any time in the previous week. The results were as follows:

    24 had not drunk any of the three

    51 had drunk tea or coffee but not a soft drink

    41 had drunk tea

    20 had drunk at least two of the three

    8 had drunk tea and a soft drink but not coffee

    9 had drunk a soft drink and coffee

    4 had drunk all three

a) Represent this information on a Venn diagram.

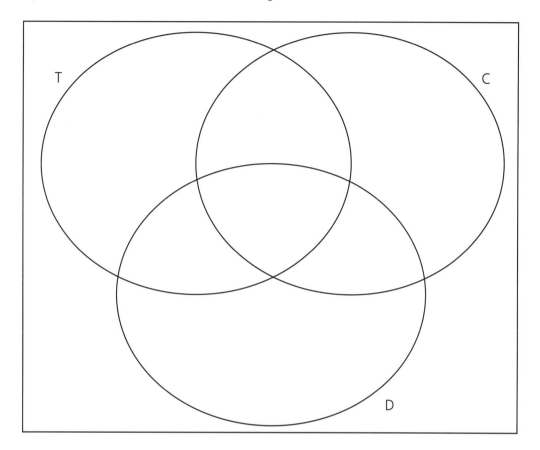

b) A student is chosen at random. Find the probability that the person picked:

   i. Had drunk tea _____

   ii. Had drunk tea or coffee but not a soft drink _____

   iii. Had drunk tea and coffee but not a soft drink _____

   iv. Had drunk soft drinks only _____

   v. Had drunk soft drinks and coffee only _____

   vi. Had drunk all three drinks _____

   vii. Had drunk at least two drinks _____

   viii. Had drunk only one drink _____

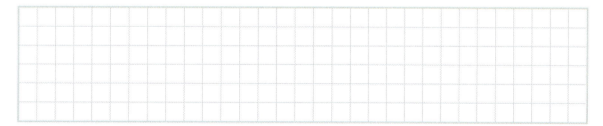

Exam-Style Questions

Exam Hints

1. Attempt every part of every question. You cannot get marks for a blank space!
2. Highlight key words in the question. Make sure you understand what you are being asked to do.
3. Write the relevant formula. The correct formula with some relevant substitution will get you marks.
4. If you are asked to give your answer in a particular format and are unsure of how to do this, do as much as you can.
5. Marks are not just given for the correct answer, they are also available for work completed along the way.

Question 1

Linked Question – Sets, Algebra, Probability, Real Numbers

35 people coming back from America were asked if they had visited Boston, New York or San Francisco. The results are as follows:

20 had visited New York

13 had visited Boston

16 had visited San Francisco

7 had visited all three

3 had visited New York and San Francisco but not Boston

1 had visited New York and Boston but not San Francisco

8 had visited Boston and San Francisco.

a) Represent this information on a Venn diagram.

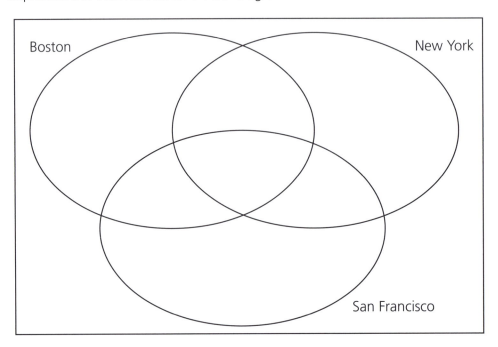

b) Let x represent the number of people who had not been to any of the three cities.
Use this information to find the value of x.

c) A person is picked at random. Calculate the probability, in its simplest form, that the person had travelled to:

i. New York only

ii. At least two cities

> At least two cities means the people who had travelled to two and three cities.

iii. None of the three cities

Question 2

Linked Question – Statistics, Probability, Real Numbers

A class of 30 students were asked how many Snapchat messages they sent in a particular week. The results are in the table below:

Number of messages	0–20	20–40	40–60	60–80	80–100
Number of students	5	9	10	4	2

CHAPTER 15: PROBABILITY

a) Represent this information on a suitable chart.

b) Using mid-interval values, calculate the mean number of Snapchat messages sent by students. Give your answer correct to the nearest whole number.

> Mid-interval value is the number that lies halfway in the interval. For example, the mid-interval value of 0–20 is 10.

c) In what interval does the median lie?

d) A student is picked at random from the group. Calculate the probability, in its simplest form, that the student chosen:

   i. Sent 60 or more Snapchats

   ii. Sent less than 20 Snapchats

e) A student is chosen from the group that sent 40 or more Snapchats. Find the probability that the student sent 60–80 Snapchats.

Question 3

Linked Question – Probability, Co-ordinate Geometry, Trigonometry, Real Numbers

a) Eugenia's soccer team are playing in a tournament. They have won 7 out of 11 of their games so far in the tournament.

Estimate the probability, to 1 decimal place, that they will win their next game.

b) Eugenia is the penalty taker. The probability that she will score is 0·87.

Calculate the probability that she will miss the next penalty she takes.

c) The points on the co-ordinate plane below represent the positions held by 3 players on the team at one stage during a match.

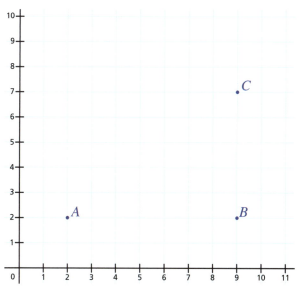

Remember: When naming a point, it's x first, then y.

Identify the points A, B and C.

$A = (\quad,\quad)$ $\qquad$ $B = (\quad,\quad)$ $\qquad$ $C = (\quad,\quad)$

d) Player B wants to make a safe pass to one of his teammates. Work out which teammate is closest to him to receive the pass.

$|BA| =$

You need to use the distance formula to calculate the distance from B to A and B to C. Don't forget to write your conclusion, i.e. which teammate is closest.

CHAPTER 15: PROBABILITY

$|BC| =$

e) Prove that *ABC* forms a right-angled triangle.

Hint: Theorem of Pythagoras:
$c^2 = a^2 + b^2$

f) Use trigonometry to find the measure of the angle $|\angle BAC|$.

$\sin \theta = \dfrac{\text{opposite}}{\text{hypotenuse}}$

$\cos \theta = \dfrac{\text{adjacent}}{\text{hypotenuse}}$

$\tan \theta = \dfrac{\text{opposite}}{\text{adjacent}}$

Summary review:

- You need to have an understanding of the terms impossible, unlikely, equally likely, likely and certain in the context of probability.

- You need to remember:
    - how to use a two-way table and a tree diagram
    - the formula for relative frequency: $\dfrac{\text{Number of times the event happens}}{\text{Total number of trials}}$
    - the formula for probability: $\dfrac{\text{Number of outcomes of interest}}{\text{Total number of outcomes}}$
    - the formula for expected frequency: Number of trials × Relative frequency or Probability.

- Independent events are events where the probability of one event occurring has no effect on the probability of another event occurring.

Scan this code to access a video with worked solutions for every question.

240   SKILLS FOR EXAM SUCCESS MATHS

Chapter 16: Geometry 2

The topics and skills I will work on are:

Congruent triangles:
- SSS
- SAS
- ASA
- RHS
- Problems involving congruent triangles

Similar triangles
- Problems involving similar triangles

Section A: Congruent Triangles

- Congruent triangles are identical triangles. The three sides and the three angles of both triangles are equal in measure.

- To prove two triangles are congruent, you do not need to prove that all three sides and all three angles are the same. Only three pieces of information are required, as outlined below.

- To prove two triangles are congruent (≡), you need to prove one of the following situations.

Side, Side, Side (SSS)

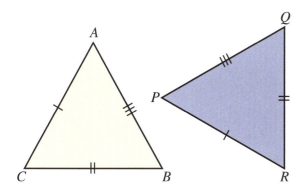

When the three sides are the same in both triangles, then the triangles are congruent.

Triangle $ABC \equiv$ Triangle PQR by SSS.

Side, Angle, Side (SAS)

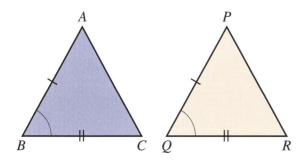

When two sides and the angle in between the two sides are equal in measure, then the triangles are congruent.

Triangle $ABC \equiv$ Triangle PQR by SAS.

Angle, Side, Angle (ASA)

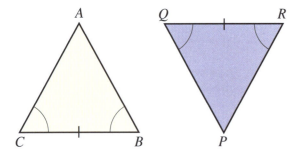

When two angles and the side in between the two angles are equal in measure, the triangles are congruent.

Triangle $ABC \equiv$ Triangle PQR by ASA.

Right angle, Hypotenuse, Side (RHS)

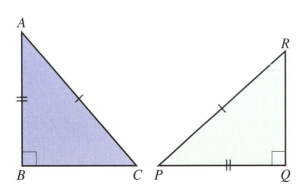

When two triangles have a side, a 90° angle and the hypotenuse the same, the triangles are congruent.

Triangle $ABC \equiv$ Triangle PQR by RHS.

Worked example

A rectangle $ABCD$ is shown below. The diagonal AC is also shown.

Prove that the Triangle ABC is congruent to the Triangle ADC.

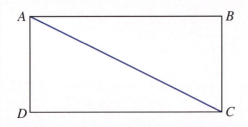

Statement	Reason
$\|AD\| = \|BC\|$	Opposite sides in a rectangle are equal in measure (S)
$\|AB\| = \|DC\|$	Opposite sides in a rectangle are equal in measure (S)
$\|AC\| = \|AC\|$	Common side (S)

△ $ABC \equiv$ △ ADC by SSS

> In the exam, you may be asked to prove that two triangles are congruent. In order to achieve full marks, you must give a reason for each statement that you make.
> For example, if you state that $\angle A = \angle B$, then you must give a reason why they are equal.

242 SKILLS FOR EXAM SUCCESS MATHS

1. The Triangle ABC is an isosceles triangle. AM is the perpendicular bisector of BC.

   Prove that the Triangle ABM is congruent to the Triangle ACM.

   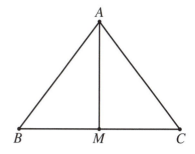

Statement	Reason
|AC| = |AB|	
	AM is the perpendicular bisector of BC
∠ABC = ∠ACM	

   △ABM ≡ △ACM by _____

2. ABCD is a parallelogram where M is the midpoint of AD.

   Prove that the Triangle ABM is congruent to the Triangle EDM.

   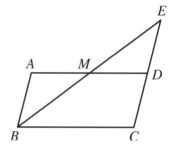

Statement	Reason
|AM| = |MD|	
	Vertically opposite angles are equal in measure
∠MAB = ∠EDM	

   △ABM ≡ △EDM by _____

3. The crossed poles in the show jump can be represented by the diagram below. |AB| and |DC| represent the two side poles and are parallel to each other.

   AC and DB are the central poles that intersect at the point E.

   Prove that the Triangle AEB and the Triangle DEC are congruent.

   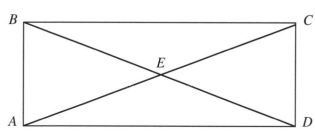

CHAPTER 16: GEOMETRY 2 **243**

Statement	Reason
$\|AB\| = \|DC\|$	
$\angle ABE = \angle EDC$	
$\angle EAB = \angle ECD$	

▲ $AEB \equiv$ ▲ DEC by _____

4. In the diagram below, PT and PS are tangents to the circle, at the points T and S.

    O is the centre of the circle, with OS and OT being radii of the circle.

    Prove that the Triangle SPO is congruent to the Triangle TPO.

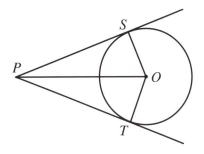

Statement	Reason
$\|SO\| = \|OT\|$	
	Where the radius meets the tangent a 90° angle is formed
$\|PO\| = \|PO\|$	

▲ $SPO \equiv$ ▲ TPO by _____

5. In the diagram below, prove that the Triangle CAO is congruent to the Triangle CBO.

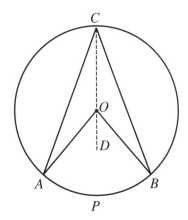

Statement	Reason

244 SKILLS FOR EXAM SUCCESS MATHS

Conclusion:

6. The diagram shows a parallelogram with vertices *A*, *B*, *C*, *D*.

   The ∠*ABC* = 120°.

   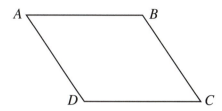

   a) Calculate the value of the following angles:

   i. ∠*ADC* =

   ii. ∠*BAD* =

   iii. ∠*BCD* =

b) Hence, or otherwise, show that the Triangles *BAD* and *BCD* are congruent.

Statement	Reason

CHAPTER 16: GEOMETRY 2   245

Conclusion:

Section B: Similar Triangles

- Two triangles are similar if they have the same shape. (This is the case for any shape, not just triangles.)

- The symbol for similarity is |||.

- To show that two triangles are similar, you must prove **one** of the following conditions.

If the lengths of matching sides are in proportion (same ratio), then the triangles are similar

> It is important that you keep the measurements consistent.
> It is also important that the triangles are in the same orientation before starting the question.

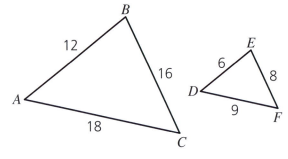

$$\frac{|AB|}{|DE|} = \frac{|CB|}{|FE|} = \frac{|AC|}{|DF|}$$

$$\frac{12}{6} = \frac{16}{8} = \frac{18}{9}$$

$$2 = 2 = 2$$

Therefore △ABC ||| △DEF

If two pairs of matching angles are equal, then the triangles are similar

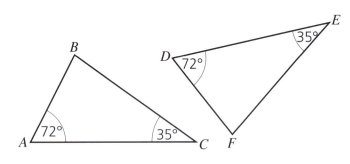

If two pairs of angles are equal, then the third pair must also be equal because the three angles in a triangle add up to 180°.

Therefore △ABC ||| △DEF

If the lengths of two sides are in proportion and the angles in between these two sides are equal, then the triangles are similar

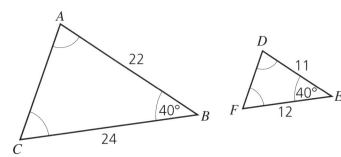

$$\frac{|AB|}{|DE|} = \frac{|BC|}{|EF|}$$

$$\frac{22}{11} = \frac{24}{12}$$

$$\frac{1}{2} = \frac{1}{2}$$

and ∠ABC = ∠DEF

Therefore △ABC ||| △DEF

In a right-angled triangle, if the length of the hypotenuse and another side are in proportion, then the triangles are similar

> This is a special case of Rule 3.

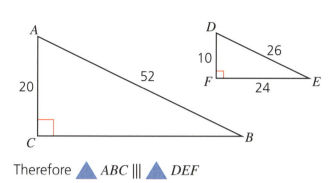

In the Triangle *ABC*, the side labelled 52 is the hypotenuse.

In the Triangle *DEF*, the side labelled 26 is the hypotenuse.

So $\dfrac{|52|}{|26|} = \dfrac{|20|}{|10|}$

$2 = 2$

Therefore △ ABC ||| △ DEF

There is a 90° angle in both triangles.

7. In the diagram below, *BC* ∥ *DE* (i.e. BC is parallel to DE). The diagram is not to scale.

   Prove that the Triangle *ABC* is similar to *ADE*.

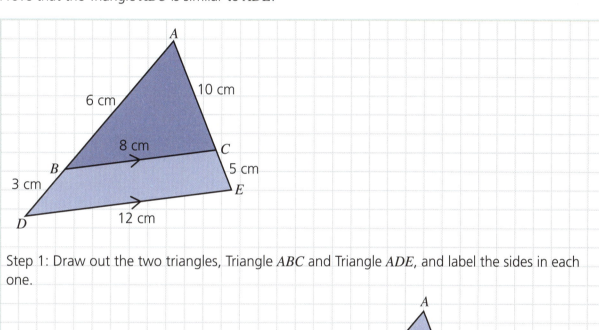

Step 1: Draw out the two triangles, Triangle *ABC* and Triangle *ADE*, and label the sides in each one.

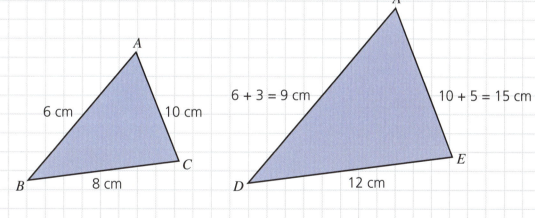

Step 2: Pair up the sides correctly.

$$\dfrac{|AB|}{|AD|} = \dfrac{|AC|}{|AE|} = \dfrac{|BC|}{|DE|}$$

Step 3: Fill the values into your ratios and conclude if the triangles are similar or not.

CHAPTER 16: GEOMETRY 2

8. In the following pairs of similar triangles, solve for the missing sides *a* and *b*.

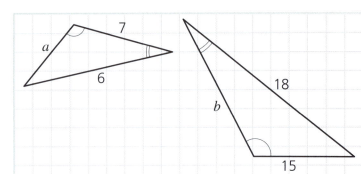

You might find it useful to redraw the triangles so that they have the same orientation. This will make it easier to see connections between the two triangles and reduce the likelihood of making mistakes.

Step 1: In this situation, the first step is to match up the sides in both triangles correctly.

As in this example, the triangles won't necessarily have the same orientation. So we need to identify the equal angles and match them with their opposite sides in each triangle.

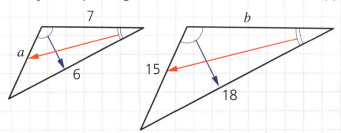

Triangle *a*	Triangle *b*
6	18
a	15
7	*b*

Step 2: Calculate *a*.

$$\frac{6}{18} = \frac{a}{15}$$

$18(a) = 6(15)$ Change the fraction into a linear equation.

$18a = 90$

$a = \frac{90}{18} = 5$

Step 3: Calculate *b*.

$$\frac{6}{18} = \frac{7}{b}$$

9. In the diagram, DE is parallel to AC.

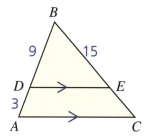

a) Prove that the Triangle BDE ||| the Triangle BAC.

Statement	Reason
$|\angle BDE| = |\angle BAC|$	
$|\angle BED| = |\angle BCA|$	
	Common angle
Conclusion:	

b) Find the length of:

i. |AB|

ii. |BC|

$$\frac{|BD|}{|BA|} = \frac{|BE|}{|BC|}$$

iii. |EC|

10. Quadrilateral *ABCD* is shown in the diagram.

    AB is parallel to *DC*.

    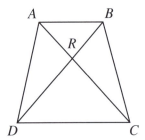

    a) Prove that the Triangle *ARB* is similar to the Triangle *DRC*.

    b) Which sides of the Triangle *ARB* correspond to the sides of the Triangle *DRC*?

    _____

    c) |*AR*| = 3, |*BR*| = 2·5, |*RC*| = 6. Find the length of |*RD*|. All values are in cm.

    d) |*AB*| = 3·5. Find |*DC*|. All values are in cm.

    > You may find it useful to draw out the individual triangles for yourself, to help see connections between the required triangles.

11. Tom wants to estimate the height of the giraffe enclosure at London Zoo.

    From the door of the enclosure, he walks 24 m. He places a 1·5 m pole in the ground and walks another 6 m.

    He notices from this point that the top of the pole and the top of the enclosure are in line. Calculate the height of the enclosure.

    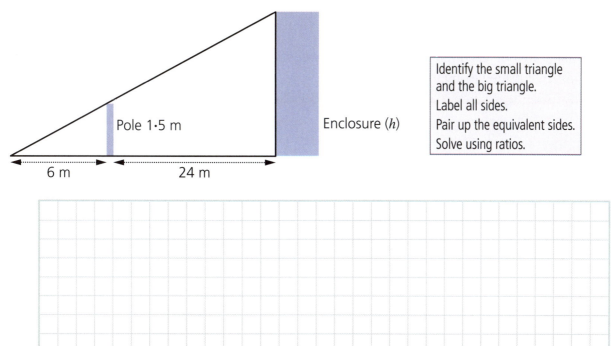

    Identify the small triangle and the big triangle.
    Label all sides.
    Pair up the equivalent sides.
    Solve using ratios.

12. Lizzie is walking along 34th Street in New York. She wants to visit the Empire State Building and stops to figure out exactly where she is.

    There is a puddle 1 m in front of her. In the puddle she sees the reflection of the top of the Empire State Building, which she knows to be 443 m high. Lizzie's eyes are 1·5 m above the ground.

    What distance is she from the Empire State Building?

    Give your answer to the nearest metre.

    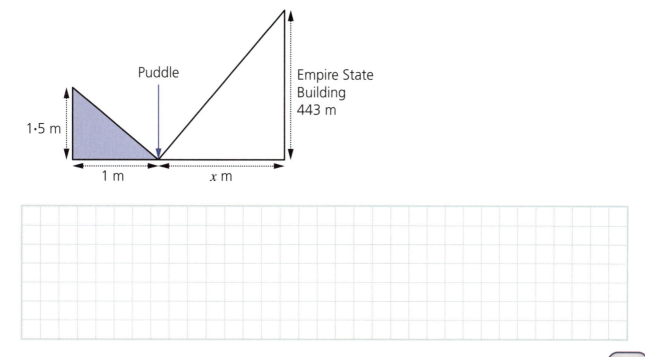

CHAPTER 16: GEOMETRY 2

Exam-Style Questions

Exam Hints

1. Attempt every part of every question. You cannot get marks for a blank space!
2. Highlight key words in the question. Make sure you understand what you are being asked to do.
3. If you are asked to give your answer in a particular format and are unsure of how to do this, do as much as you can.
4. Marks are not just given for the correct answer, they are also available for work completed along the way.

Question 1

Linked Question – Geometry, Co-ordinate Geometry, Constructions, Real Numbers

a) On the co-ordinate plane, plot and label the following points:

   $A\,(2, 2)$ $B\,(6, 7)$ $C\,(10, 2)$

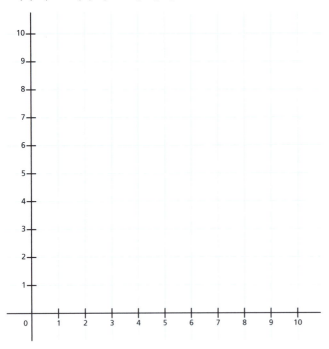

b) Using the distance formula, calculate the length of $|AB|$ and $|BC|$.

c) What type of triangle is the Triangle *ABC*? How do you know?

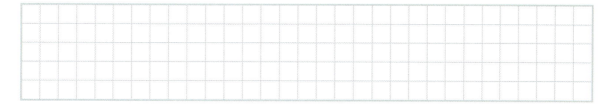

d) Using a compass and a ruler, construct the perpendicular bisector of *AC*.
   Show all construction lines clearly.

   Where the perpendicular bisector cuts the line *AB*, mark this as point *D* on your diagram.

e) Prove that the Triangle *ABD* is congruent to the Triangle *BDC*.

f) Use trigonometry to find the value of |∠*BAD*|.

   Give your answer correct to the nearest degree.

Question 2

Linked Question – Geometry, Trigonometry, Real Numbers, Distance/Time/Speed

Due to the high volume of entries for a charity race, it is decided to set up two courses for competitors.

Competitors will have to swim a given distance, cycle a given distance and run a given distance. The distances for swimming, cycling and running may not be the same in both courses.

a) This diagram represents Course 1.

   Contestants will have to swim for 5 km and run for 12 km.

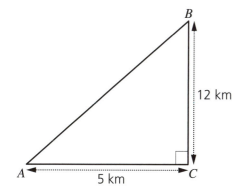

Remember: You are dealing with right-angled triangles so:

$\sin \theta = \dfrac{\text{opposite}}{\text{hypotenuse}}$ $\cos \theta = \dfrac{\text{adjacent}}{\text{hypotenuse}}$

$\tan \theta = \dfrac{\text{opposite}}{\text{adjacent}}$ $c^2 = a^2 + b^2$

CHAPTER 16: GEOMETRY 2

i. Use the theorem of Pythagoras to calculate the distance that contestants will have to cycle.

ii. Find |∠BAC| correct to 1 decimal place.

iii. Hence, or otherwise, find |∠ABC|.

b) This diagram represents Course 2.

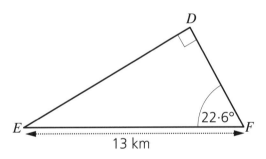

i. DF represents the distance that contestants will have to swim.

Find, correct to the nearest whole number, the distance contestants will have to swim.

ii. Calculate the length of |DE|.

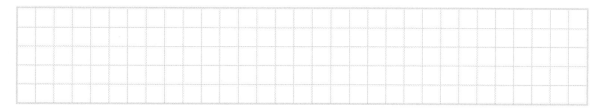

iii. Find |∠DEF|, correct to 1 decimal place.

c) Prove that the total length of the two courses is identical.

d) Donagh completes the run in Course 1 in a time of 58 minutes 32 seconds.

Calculate his speed in **metres per second**, correct to the nearest whole number.

Remember:
1 km = 1000 m
1 minute = 60 seconds

Question 3

Linked Question – Geometry, Algebra

In the diagram below, AB is parallel to DE.

$|\angle BAC| = (x + y)°$, $|\angle ABC| = 60°$, $|\angle ACB| = (3x + y)°$ and $|\angle DCE| = 70°$.

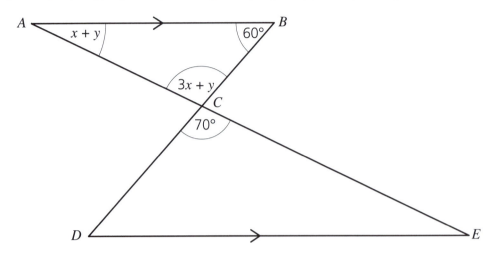

a) Form two equations in x and y and use simultaneous equations to find the value of x and the value of y.

> Remember: The three angles in a triangle add up to 180°. Vertically opposite angles are equal in measure.

b) Hence, or otherwise, mark in the values of each of the angles in the diagram. Give a reason for each answer.

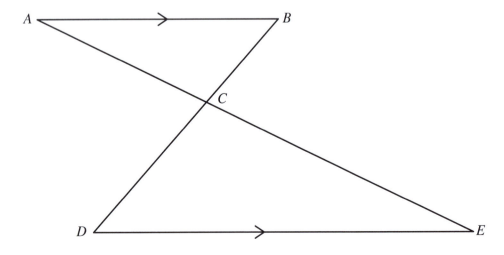

Angle	Reason
$|\angle BAC| =$	
$|\angle ABC| =$	
$|\angle ACB| =$	
$|\angle DCE| =$	
$|\angle CDE| =$	
$|\angle CED| =$	

c) What can you conclude about the two triangles?

d) Hence, or otherwise, find the value of a and b on the diagram.

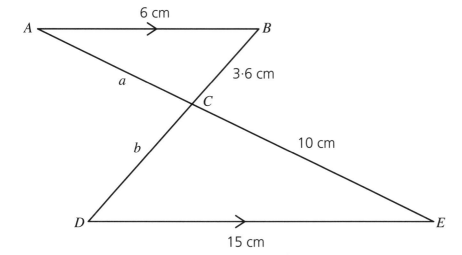

It may be useful to draw the triangles out in the same orientation before assigning sides to the formula. This will help reduce the chance of making a mistake.

Summary review:

- To prove two triangles are congruent (≡), one of the following conditions must be met: SSS, SAS, ASA or RHS.

- When trying to find a missing side in similar triangles, the following is vital: If you put the measurement from the first triangle above the line in the first ratio, then the measurement from the first triangle must go above the line in each of the subsequent ratios.

Scan this code to access a video with worked solutions for every question.

Chapter 17: Geometry 3

> To complete the constructions, you will need the following pieces of equipment: pencil, rubber, compass, ruler, protractor and set square.

The topics and skills I will work on are:

How to bisect an angle ☐

Perpendicular bisector of a line segment ☐

Drawing a line perpendicular to a given line l, passing through a given point not on l ☐

Drawing a line perpendicular to a given line l, passing through a point on l ☐

Drawing a line parallel to a given line l, through a point not on the line l ☐

Dividing a line segment into equal parts ☐

Constructing a triangle given:
- the length of the three sides ☐
- the length of two sides and the angle between the two sides ☐
- the length of one side and the measure of the two angles at either end of the given side ☐
- A right angle, the length of the hypotenuse and the length of one other side ☐

Constructing a parallelogram ☐

Transformations
- Translations ☐
- Axial symmetry ☐
- Central symmetry ☐
- Axis of symmetry ☐
- Centre of symmetry ☐

Section A: How to Bisect an Angle

Exam Hint

When carrying out a construction, it is **vital** that you show all construction lines. When the examiner is assessing your work, marks will be deducted if the construction lines are not there.

Scan this code to access a video of this construction.

1. Use your protractor to construct an angle of 70°.

   Bisect the angle. Show all construction lines.

CHAPTER 17: GEOMETRY 3 259

2. Use your protractor to construct an angle of 110°.

   Bisect the angle. Show all construction lines.

   Remember!
   When reading measurements from a ruler or a protractor, read from above and not from the side to avoid the error of parallax.

   Scan this code to access a video of this construction.

Section B: Perpendicular Bisector of a Line Segment

Scan this code to access a video of this construction.

3. Using your ruler, draw a line that is 10 cm in length.

   Construct the perpendicular bisector of this line.

   Show all construction lines clearly.

   Exam Hint
   When carrying out your constructions, it is important to be as accurate as possible.

   There is a margin of error/tolerance built into the marking scheme, usually ± 2 mm or ± 2°.

4. Using your ruler, draw a line that is 14 cm in length.

   Construct the perpendicular bisector of this line.

   Show all construction lines clearly.

   Scan this code to access a video of this construction.

260 SKILLS FOR EXAM SUCCESS MATHS

Section C: Drawing a Line Perpendicular to a Given Line l, Passing Through a Given Point Not on l

5. In each of the following, draw a line through the given point, perpendicular to the line.

   a)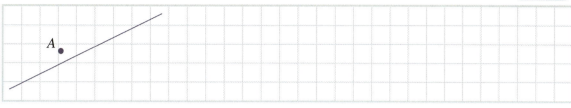

   b)

Section D: Drawing a Line Perpendicular to a Given Line l, Passing Through a Point on l

6. In each of the following, draw a line through the given point, perpendicular to the line that contains the point.

   a)

   b)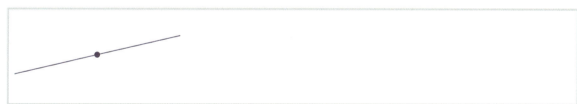

Section E: Drawing a Line Parallel to a Given Line l, Through a Point Not on the Line l

7. In each of the following, draw a line through the given point, parallel to the given line.

   a)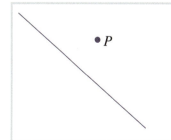

CHAPTER 17: GEOMETRY 3 | **261**

b)

Section F: Dividing a Line Segment into Equal Parts

8. Using a ruler, draw a line that is 12 cm in length.

   Divide the line segment into 3 equal parts, clearly showing **all** construction lines.

9. Using a ruler, draw a line that is 18 cm in length.

   Divide the line into 4 equal parts, clearly showing **all** construction lines.

262 SKILLS FOR EXAM SUCCESS MATHS

Section G: Constructing a Triangle Given the Length of the Three Sides

10. a) Construct the triangle ABC where |AB| = 7 cm, |BC| = 5 cm and |AC| = 6 cm.

    Show all construction lines clearly.

    b) Using your compass and a straight edge, bisect |∠ABC|.

    Show all construction lines clearly.

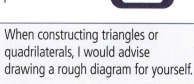

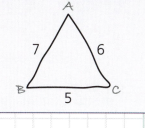

11. a) Construct the Triangle ABC where |AB| = 8 cm, |BC| = 6 cm and |AC| = 10 cm.

    Show all construction lines clearly.

    b) Construct the perpendicular bisector of AC.

    Show all construction lines clearly.

Section H: Constructing a Triangle Given the Length of Two Sides and the Angle Between the Two Sides

12. a) Construct the Triangle ABC where |AB| = 6 cm, |BC| = 7 cm and |∠ABC| = 60°.

    Show all construction lines clearly.

CHAPTER 17: GEOMETRY 3 | 263

b) Through the point *A*, construct a line parallel to *BC*.

13. Construct the Triangle *ABC* where |*AB*| = 5 cm, |*BC*| = 4 cm and |∠*ABC*| = 100°.

    Show all construction lines clearly.

Scan this code to access a video of this construction.

Section I: Constructing a Triangle Given the Length of One Side and the Measure of the Two Angles at Either End of the Given Side

14. **a)** Construct the Triangle *ABC* where |*AB*| = 7 cm, |∠*CAB*| = 50° and |∠*CBA*| = 50°.

    b) Construct the bisector of |∠*ACB*|. Extend the bisector to the side of *AB* and label this point *D*.

    c) Prove that the Triangle *ACD* is congruent to the Triangle *BCD*.

Scan this code to access a video of this construction.

264   SKILLS FOR EXAM SUCCESS MATHS

15. a) Construct the triangle *ABC* where |*AB*| = 6 cm, |∠*CAB*| = 50° and |∠*CBA*| = 70°.

    b) Construct the bisector of |∠*CAB*| and extend the bisector to the side of *BC*. Mark this point *D*.

    c) Is the Triangle *ACD* congruent to the Triangle *CBD*?

    Explain your answer.

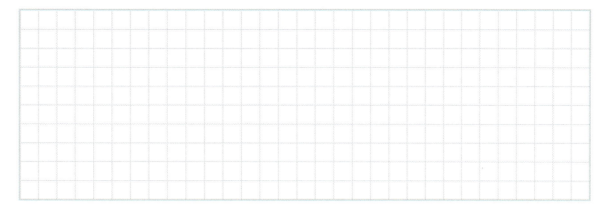

Section J: Constructing a Triangle Given a Right Angle, the Length of the Hypotenuse and the Length of One Other Side

16. Construct a Triangle *ABC* such that |*BC*| = 4 cm, |∠*ABC*| = 90° and |*AC*| = 5 cm.

    Show all construction lines clearly.

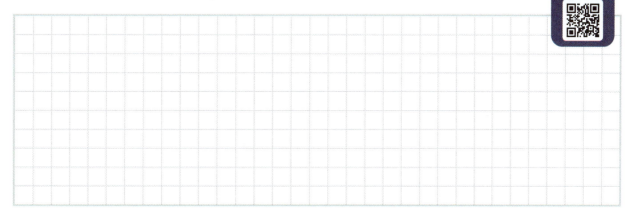

17. a) Construct a Triangle *ABC* such that |*BC*| = 5 cm, |∠*ABC*| = 90° and |*AC*| = 13 cm.

    b) Construct the bisector of |∠*ABC*|. Extend the bisector to the side of *AC* and label it point *D*.

    c) Are the Triangles *ABD* and *BCD* similar to each other?

    Explain your answer.

Section K: Constructing a Parallelogram

Scan this code to access a video of this construction.

18. a) Construct a Parallelogram such that |AB| = 6 cm, |AD| = 4 cm and |∠ADC| = 50°.

    b) Construct the diagonal BD.

    c) Are the Triangles ABD and BCD congruent to each other?

    Explain your answer.

19. a) Construct a parallelogram such that |AB| = 8 cm, |BC| = 11 cm and |∠BCD| = 70°.

    b) Construct the diagonal AC.

    c) Are the Triangles ABC and ADC similar?

    Explain your answer.

Section L: Transformations

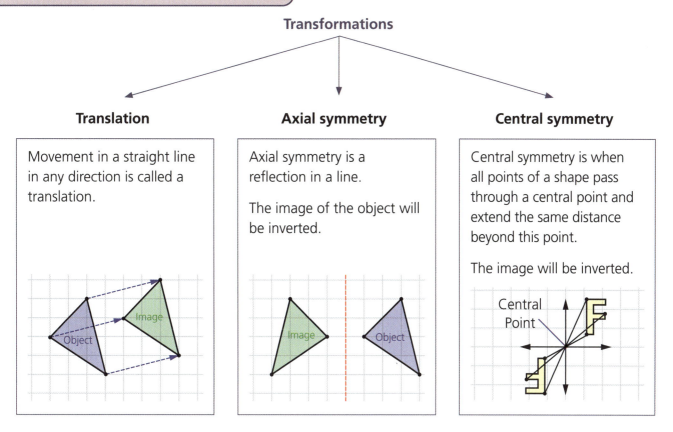

20. Write A, B, C, D in the appropriate places in the table below to show which image comes from each transformation. You can use each letter only once.

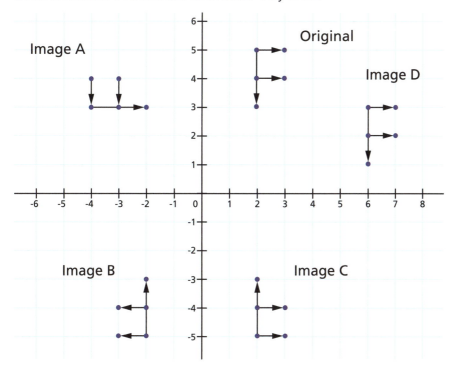

Transformation	Image (A, B, C, D)
Central symmetry through (0, 0)	
Translation	
Rotation	
Axial symmetry in the *x*-axis	

CHAPTER 17: GEOMETRY 3

21. Draw the image of the given triangle through an axial symmetry in the given line.

Scan this code to access a video of this construction.

22. Draw the image of the given shape through a central symmetry in the given point.

Scan this code to access a video of this construction.

23. How many axes of symmetry do the following shapes have?

Draw each axis on the given diagrams.

A shape has an axis or line of symmetry when one half of the shape fits exactly over the other half when the shape is folded along the line.

a)

Number of axes:
_____

b)

Number of axes:
_____

c)

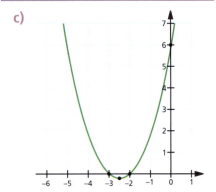

Number of axes:
_____

268 SKILLS FOR EXAM SUCCESS MATHS

Exam-Style Questions

Exam Hints

1. Attempt every part of every question. You cannot get marks for a blank space!
2. Highlight key words in the question. Make sure you understand what you are being asked to do.
3. If you are asked to give your answer in a particular format and are unsure of how to do this, do as much as you can.
4. Marks are not just given for the correct answer, they are also available for work completed along the way.

Question 1

Linked Question – Geometry, Algebra, Real Numbers, Constructions

In the Parallelogram $ABCD$, $|\angle A| = 2x + 6y°$, $|\angle B| = 70°$, $|\angle C| = 110°$ and $|\angle D| = 4x + 2y$ as shown in the diagram.

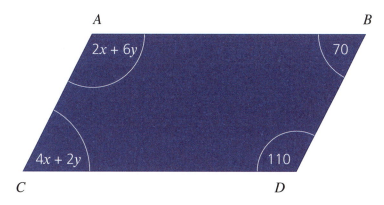

a) Use simultaneous equations to find the value of x and y.

> Opposite angles in a parallelogram are equal in measure.

b) Hence, or otherwise, find the value of each of the angles in the parallelogram.

CHAPTER 17: GEOMETRY 3 **269**

c) The perimeter of the parallelogram is 36 cm.

The ratio of |AB| : |AC| is 5 : 4.

Calculate the length of:

i. |AB|

ii. |AC|

> Opposite sides in a parallelogram are equal in measure.

d) Construct the Parallelogram ABCD where |AB| = 8 cm, |BC| = 6 cm and |∠ABC| = 130°.

Scan this code to access a video of this construction.

Question 2

Linked Question – Co-ordinate Geometry, Transformations

a) On the co-ordinate plane, construct the Triangle ABC where A (3, 3), B (7, 1) and C (9, 5).

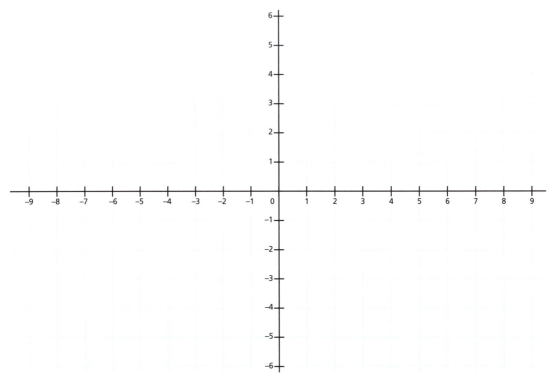

b) Prove that *AB* is perpendicular to *BC*.

Hint:
Find the slope of *AB*.
Find the slope of *BC*.
If lines are perpendicular,
slope *AB* × slope *BC* = −1.

c) Draw an image of the Triangle *ABC* under axial symmetry in the *y*-axis.

Label the image *DEF*.

d) Use the theorem of Pythagoras to prove that the Triangle *DEF* is also right-angled.

Hint: Distance formula to find the length of each side.

e) Hence, or otherwise, prove that the Triangles *ABC* and *DEF* are congruent.

Criteria for congruency:
SSS
SAS
ASA
RHS

Scan this code to access a video of this construction.

Question 3

Linked Question – Constructions, Geometry, Real Numbers, Trigonometry

a) Construct the Triangle *ABC* such that $|AB| = 10$ cm and $|AC| = |BC| = 8$ cm.
   Show all construction lines clearly.

CHAPTER 17: GEOMETRY 3

b) Construct the perpendicular bisector of *AB*. Where the bisector cuts through the line |*AB*|, label it point *D*. Extend the line so that it passes through the vertex *C*, creating the triangles *ADC* and *BDC*.

c) Use the theorem of Pythagoras to find the length of |*CD*|. Give your answer in surd form.

Exam Hint

Even if you are unsure of how to give your answer in the required format, do as much of the question as you can. There are marks available for each correct step along the way.

d) Prove that the perpendicular bisector of *AB* also bisects the angle |∠*ACB*|.

Hint:
In right-angled triangles:
$\tan \theta = \dfrac{\text{opposite}}{\text{adjacent}}$
$\cos \theta = \dfrac{\text{adjacent}}{\text{hypotenuse}}$
$\sin \theta = \dfrac{\text{opposite}}{\text{hypotenuse}}$
Here you are trying to show that |∠*ACD*| = |∠*BCD*|.

e) Hence, or otherwise, prove that the Triangles *ADC* and *BDC* are congruent.

Scan this code to access a video of this construction.

Summary review:

- You need to remember each of the steps involved in each of the constructions.
- You must remember to show all construction lines clearly.
- You need to remember how to find the image of a point through both an axial symmetry and a central symmetry.

272 SKILLS FOR EXAM SUCCESS MATHS

Answers

Chapter 1: Numbers

1. a) 3, 5   b) 4, 9, 10, 12, 15, 18, 30
   c) 12, 18, 30   d) 4, 5, 10
   e) 5, 10, 15, 30
2. 2, 3, 5, 7, 13
3. She needs 3 packets of burgers and 2 packets of buns.
4. a) 9   b) 4   c) −3   d) −13
5. a) 3   b) −8   c) 12   d) −20   e) 4
   f) −27
6. a) $\frac{1}{2}$ b) $\frac{4}{5}$ c) $\frac{2}{7}$
7. $\frac{17}{20}$
8. a) $15\frac{2}{5}$ b) $\frac{5}{8}$
9. 2400 g
10. a) 5·78   b) 8·977   c) 127·9879
11. a) 0·00238   b) 0·000654
    c) 19 700   d) 22 300
12. π, as it cannot be written as a fraction
13. $5\sqrt{3}$
14. −36
15. a) a^7 b) $2^4 = 16$
16. 5
17. $7·8 \times 10^8$

Exam-Style Questions

Question 1

a)

Number	$\mathbb{N}$	$\mathbb{Z}$	$\mathbb{Q}$	$\mathbb{Q}/\mathbb{R}$	$\mathbb{R}$
$\sqrt{3}$	No	No	No	Yes	Yes
3	Yes	Yes	Yes	No	Yes
−3	No	Yes	Yes	No	Yes
$\frac{2\pi}{3}$	No	No	No	Yes	Yes
0·33	No	No	Yes	No	Yes

b) i. 72 students   ii. 35%   iii. 50 girls

Question 2

a) i. $\frac{3}{5} + \frac{9}{25} = \frac{24}{25}$ ii. $\frac{25}{3} - \frac{5}{9} = \frac{70}{9}$
b) 27

Question 3

$5\sqrt{2}$

Question 4

$-4 = \mathbb{Z}$; $\pi = \mathbb{R}$; $\frac{5}{6} = \mathbb{Q}$; $\sqrt{11} = \mathbb{R}$;
$0·45 = \mathbb{Q}$; $4^{-2} = \mathbb{Q}$

Chapter 2: Applied Arithmetic

1. €144·99
2. €3000
3. €41 666·67
4. €66·11
5. 13·5%
6. €68·75
7. I recommend that he buy the vest from the American website as it is cheaper.
8. €1 = 4·69 złoty
9. a) €18 000   b) €13 800
   c) €1987·38   d) €1800
   e) €2860   f) €39 552·62
10. €62 333
11. €6900
12. €18 693
13. €31 722·60
14. €11 278·30
15. 2·5%
16. €3000

Exam-Style Questions

Question 1

a) €656·10   b) i. 20% of €44 300 = €8860; 40% of €11 700 = €4680
ii. €45 760

Question 2

a) Special Offer A   b) €6·25
c) €3311·44

Question 3

a) Sinead = €560, Veronika = €640
b) €581   c) $697·60   d) €48·78
e) 6 hours 10 minutes   f) 183 km

Chapter 3: Sets

1. a) Venn diagram
   b) i. True   ii. False   iii. True
      iv. False   v. True   vi. False
   c) i. {a, b, c, d, e, f, g, h, i}   ii. {d, e}
      iii. {f, g, h, i, j, k}   iv. {a, b, c, j, k}
      v. {a, b, c}   vi. {f, g, h, i}
      vii. {j, k}   viii. {a, b, c, d, e, f, g, h, i, j, k}   ix. {f, g, h, i, j, k}
      x. {a, b, c, f, g, h, i, j, k}
4. b) i. 15   ii. 11   iii. 4   iv. 26
6. a) 6   b) Diagram
   c) i. 10   ii. 6   iii. 26
7. 2
8. a) 3   b) 5   c) 4   d) 7   e) 26
   f) 12   g) 11   h) 4
9. b) 12   c) i. $\frac{1}{10}$ ii. $\frac{21}{40}$ iii. $\frac{9}{40}$
   iv. $\frac{19}{40}$
10. b) 2

Exam-Style Questions

Question 1

54

Question 2

b) ii. $\#P \leq \#Q$ or $\#P = \#Q$; P is a subset of Q so all the elements in Set P must also be in Set Q

Question 3

b) 3   c) i. $\frac{4}{11}$ ii. $\frac{4}{11}$ iii. $\frac{3}{11}$
d) 15 : 11   e) Histogram
f) $8·05 \times 10^5$

Chapter 4: Algebra 1

1. a) −5   b) 3   c) $2\sqrt{5}$
2. a) €280   b) €275   c) 1·76%, No she is not correct
3. a) $10x + 18$ b) $3x^2 - 9x - 2$
   c) $-6x^3 - 13x^2 + 6x + 8$
   d) $2 - 11x + 8x^2 - 5x^3$
4. a) $-5x^3 - 6x^2 + 4x - 13$
   b) $-10 + 9x - 2x^2 + 3x^3$
5. a) $x^3 + x^2 - 8$ b) $9x^3 - 8x^2 - 3x$
6. a) $6x^2 + x - 12$ b) $-12x^2 - 5x + 3$
   c) $2x^3 - x^2 - 8x - 5$
   d) $6x^4 - 22x^3 + 38x^2 - 33x + 5$
   e) $16x^2 - 40x + 25$
   f) $27x^3 - 54x^2 + 36x - 8$
7. a) Paul = x b) Ellie = $x - 5$
   c) Sophie = $x - 3$ d) Jack = $4x - 12$
   e) All ages = $7x - 20$
8. $\frac{6x + 31}{20}$
9. $\frac{6x - 26}{12}$
10. $\frac{11x - 8}{(2x - 1)(3x - 2)}$
11. $\frac{-2x - 16}{(5 + 4x)(2x - 2)}$

Exam-Style Questions

Question 1

a) 36   b) $\frac{7}{(2x + 1)(3x + 5)}$ c) 36°

Question 2

a) −6   b) $\frac{20 - 2x^2 - x}{(2x + 1)(4)}$

Question 3

a) $3x^3 - 14x^2 + 23x - 10$
b) $\frac{2y + 23}{(y + 2)(2y + 5)}$ c) Jane's bottle volume = 1571 cm³, Roisín's bottle volume = 1256·8 cm³; Roisín is incorrect
d) Cylinder 1 = $2\pi r^3$, Cylinder 2 = $3\pi r^3$.

Chapter 5: Algebra 2

1. a) 2   b) 7   c) i. $x + 6$ ii. 30 m
   iii. 1080 m²
2. a) Points to be plotted: (0, 10), (1, 10.5), (2, 11), (3, 11·5), (4, 12), (5, 12·5), (6, 13), (7, 13·5)
   d) 15 km   e) 32 days
3. $\frac{9}{11}$
4. $\frac{3}{4}$
5. $\frac{1}{3}(x + 6) - \frac{1}{4}(x)$, 36, 42
6. $x = 15, y = \frac{-3}{2}$
7. $x = 6, y = 4$
8. a) $x = €15, y = €20$ b) €126
9. x (adults) = 125, y (children) = 175
10. (4, 1)
11. {−4, −3, −2, −1, 0, 1, 2, 3, 4, 5}
12. $u = \sqrt{v^2 - 2s}$
13. a) $\cos A = \frac{a^2 - b^2 - c^2}{-2bc}$
    b) $A = \cos^{-1}\left(\frac{a^2 - b^2 - c^2}{-2bc}\right)$

ANSWERS 273

Exam-Style Questions
Question 1
b) $\frac{-3}{2}$ b) 5   c) $E = 30, F = 20$
d) 1·9 cm

Question 2
a) i. Natural numbers are positive whole numbers   ii. Integers are positive and negative whole numbers   c) $5 > y \geq -1$

Question 3
a) €35 905·71   b) 6 m   c) €3699
d) 10·3%   e) i. $8 + 3$ m   ii. 32 cm
iii. Yes, as the difference between each term is constant.

Chapter 6: Algebra 3
1. a) $5x(x + 2)$ b) $3ab(1 + 2b - 3a)$
   c) $2x(-x^2 + 2x - 3)$
   d) $xy(1 - 5x + 6y)$
2. a) $(a + c)(a - b)$ b) $(7 + 3x)(p - q)$
   c) $(x - 2y)(x + 3)$
3. a) $(10y - 15x)(10y + 15x)$
   b) $(y - 1)(y + 1)$
   c) $(3a - 6b)(3a + 6b)$
   d) $(7 - p)(7 + p)$
4. a) $(x - 2)(x + 3)$ b) $(x - 7)(x - 4)$
   c) $(x - 8)(x + 6)$ d) $(x + 13)(x + 7)$
5. a) $(2x - 7)(3x + 4)$
   b) $(5x + 1)(3x - 2)$
   c) $(4x - 3)(3x - 5)$
6. a) $x = \frac{3}{4}, x = -3$ b) $x = \frac{-4}{3}, x = 3$
   c) $x = \frac{2}{5}, x = -1$
7. $\frac{2f - 5}{f - 3}$
8. $\frac{2x}{2x - 3}$
9. $3 \pm 4\sqrt{2}$
10. $x = 4$ and $x = -1$, so the natural numbers are 4, 5, 6
11. $x = 25, x = 1$, the value cannot be $x = 1$
12. $3x^2 + 6x + 5$
13. Divides evenly so $x - 2$ is a factor. The other two factors are $x = -\frac{5}{4}$ and $x = \frac{3}{2}$

Exam-Style Questions
Question 1
a) 2·5 cm   b) i. $4x + 4$ ii. 5
c) i. Triangle A = 8 cm, Triangle B = 24 cm, Triangle C = 40 cm
ii. $y = 4$

Question 2
a) $(h + 3)(5f - 2h)$ b) $\frac{2 \pm 2\sqrt{3}}{4}$ c) $\frac{3x + 7}{x - 5}$
d) $5x^2 - 3x + 2$

Question 3
a) Length: $136 - 2x$; Width: $86 - 2x$; $x = 108$ and $x = 3$. x cannot be 108 m so $x = 3$ m   b) 1296 m²   c) €97 200
d) €6304   e) 12°

Question 4
a) Graph 1: $f(x)$, Graph 2: $g(x)$
b) $f(x) = x^2 - x - 6 = 0$,
$k(x) = x^2 + x - 6 = 0$

Chapter 7: Co-ordinate Geometry of the Line
2. $(8, -3)$
3. $3\sqrt{10}$
4. $|AB| = 3, |CD| = \frac{1}{7}$
5. $3x - y - 11 = 0$
6. The point (1, 4) is not on the line; the point (2, 5) is not on the line
7. a) Equation of the line: $y = \frac{2}{3}x - 6$
   b) Slope $(m) = \frac{3}{2}$, y-intercept $= -6$
9. a) Point of intersection: $(-3, 7)$
   c) Solving with simultaneous equations is more accurate as it gives the exact solution rather than trying to estimate from a diagram.
10.

Equation	y-intercept	Slope
$y = x + 3$	3	1
$y = x - 3$	-3	1
$y = 2x + 3$	3	2
$y = 2x - 3$	-3	2
Parallel?	Perpendicular?	Line
t	no	r
r	no	t
s	no	q
q	no	s

11. a) $2 \times \frac{1}{-2} = -1$, therefore the lines are perpendicular to each other.
    b) $P(0, -5)$ c) $K(0, \frac{23}{2})$
12. $x - 3y - 7 = 0$

Exam-Style Questions
Question 1
a) $P = (-1, 3($ b) $(0, \frac{20}{7})$ c) $7x - y - 40 = 0$ d) 1·42 km

Question 2
$(0, -68)$

Question 3
a) $B = (8, 0), C = (10, 4)$ b) $|AC|$: Slope $(m) = \frac{1}{3}$, $|AB|$: $k = 4$ c) 10 units²
d) Diagram   e) Point B   f) $(p, -s)$
g) $40 = 40$, therefore triangle ABC is right angled   h) Point of intersection $= (4, 2)$

Chapter 8: Number Patterns
1. a) 60, 80, 100, 120, 140, 170
   c) €150   d) $T_n = 10n + 20$
   e) i. €270   ii. €250   f) 58 weeks

g) Slope = 10 means that the money is increasing by €10 per week
h) The point (20, 220) is on the line
i) After 20 weeks, €220 is saved
2. b) 24·5 m   c) 2·5 seconds
   d) Quadratic, the second difference is constant   f) $-n^2 + 12n - 11 = 0$

3. a)

0	1	2	3	4	5	6
1	2	4	8	16	32	64

c) Exponential, each term is generated by doubling the previous term
d) $1·05 \times 10^6$

Exam-Style Questions
Question 1
a)

Distance	1	2	3	4	5	6	8	10
A	5	6	7	8	9	10	12	14
B	2	4	6	8	10	12	14	16

c) Yes, the difference between each term is constant
d) Company A, cheaper   e) €56
f) 45 minutes   g) €6·45   h) (4, 8)

Question 2
b)

Pattern no.	1	2	3	4	5	6
No. of matchsticks	3	7	11	15	19	23

c) Linear, the difference between each term is constant.   d) $4n - 1$ e) 71
f) $95\sqrt{3}$

Question 3
a) i. 43, 55, 69   ii. Second difference is constant so quadratic sequence
c) $b = -1, c = 13$ d) 103

Chapter 9: Functions and Graphing Functions
1. a) Function, one output for each input; b) Not a function, 2 generates two outputs; c) Function, one input for each output
2. a) ii. -12 iii. -15 iv. -16
   v. -15 c) $x = 5, x = -3$
3. b) $(1, -1)$ c) $(1, -1)$
4. b) i. $x = -1·5, x = 2$
   ii. $x = -1·8, x = 2·2$ iii. $y = -5·3$
   iv. Minimum value $= -6$
   v. Minimum point $= (0, -6($
   vi. $f(x($ increasing: $0 \leq x \leq 3$;
   $f(x($ decreasing: $-2 \leq x \leq 0$
   vii. $(-1, -2)$ and $(2, 1)$
   viii. $f(x) \leq g(x)$ when $-1 \leq x \leq 2$
5. a) i. and ii.

Solve $f(x) = 0$
i. $2x^2 + x - 6$
$x = \frac{3}{2}$ and $x = -2$

ii. Identify where $f(x)$ cuts the y-axis
$y = -6$
Solve $g(x) = 0$
$x^2 - 6x + 9$
$x = 3$
Identify where $g(x)$ cuts the y-axis
$y = -9$
Solve $h(x) = 0$
$x^2 - 2x$
$x = 0$ and $x = 2$
Identify where $h(x)$ cuts the y-axis
Does not cut the y-axis

b) Diagram 3 = $f(x)$, Diagram 5 = $g(x)$, Diagram 2 = $h(x)$
6. (0, 3), (1, 6), (2, 12), (3, 24)
7. b) Cellulon starts from (0, 0) therefore has no fixed charge c) (500, 2000) d) If she uses less than 500 MB, she should go with Cellulon. If she uses more than 500 MB, she should go with Mobil. The point of intersection shows us the point where the cost of Mobil exceeds that of Cellulon
8. a) 5 m/s b) 2·5 m/s
   c) Claire, Bill, Dee

Exam-Style Questions
Question 1
a) Lecky, graph starts from (0, 0)
b) Domain $0 \leq x \leq 1000$, Range $50 \leq b(x) \leq 325$ c) i. (100, 70) and (800, 280) ii. $100 < x < 800$
iii. Buzz is cheaper if the number of units used is between 100 and 800
d) i. $\frac{11}{40}$ ii. The cost of electricity rises by $\frac{11}{40}$ for every one unit increase of usage e) i. (0, 50) ii. $y = \frac{11}{40}x + 50$

Question 2
a) (0, 130), (1, 90), (2, 70), (3, 70), (4, 90), (5, 130) b) i. 130 m ii. 70 m
iii. 2·5 hours c) i. $c = 180$ ii. $a = 20$, $b = -120$

Question 3
b) 8, 14, 20 c) The second difference is not constant so it is not a quadratic sequence d) 11·4 km/hr e) Ella's speed decreases. The first 2 kilometres took 8 minutes and the next 2 kilometres took 12 minutes f) Ciarán stopped running for 5 minutes

Chapter 10: Trigonometry
1. a) 13 b) 15
2. 8 cm
3. a) 63·44 b) 26·96
4. $\sin \theta = \frac{5}{13}$ $\cos \theta = \frac{5}{12}$
5. a) 75° b) 23°
6. a) 29·05° b. 64·42°
7. 65°
8. a) 16·48

9. 26·9°
10. 2·34 km
11. a) 36·628 m b) 103·6 m
    c) 87·72 m

Exam-Style Questions
Question 1
a) 12·06 b) 11·37° c) 72 km/hr
d) i. 1·2 m ii. 1·68 m e) $t = 0.5$ and $t = 2$ seconds f) (0·5, 3·2) (1, 4·2) (1·5, 4·2) (2, 3·2) (2·5, 1·2)
g) i. 4·4 m ii. 2·7 seconds
iii. Axis of symmetry on graph

Question 2
a) 55° b) 269 m c) 56 m

Question 3
b) i. Isosceles ii. Two sides are equal in length d) 3·6 cm e) 45·57°
f) Congruent by ASA

Question 4
a) ii. $\frac{-3}{4}$ b) i. $|BC| = 3$ ii. $|AB| = 5$
c) i. $\tan A = \frac{3}{4}$ ii. The slope of the line and the tan of the angle are equal
d) $A' = (5, -2)$, $B' = (1, -2)$, $C' = (1, -5)$

Chapter 11: Geometry 1
1.
a) A theorem is a statement that may be proved by following a certain number of logical steps.
b) A proof is evidence which shows the truth of a given statement.
c) An axiom is a rule/statement we accept as true without any proof. Example: a straight line may be drawn between any two points.
d) A corollary is a statement that follows readily from a previous theorem. Example: each angle in a semi-circle is a right angle.
e) A converse is formed by swapping the order of the hypothesis and the conclusion. Example: Statement: In an isosceles triangle, the angles opposite the equal sides are equal in measure. Converse: If two angles in a triangle are equal in measure, it is an isosceles triangle.
2. A and C, B and D, F and H or G and E b) B and F or C and G c) A and G or D and F d) A and F or D and G e) i. and ii.

Angle	Measure	Reason
B	43°	180° − 137°
C	137°	Vertically opposite to A
D	43°	Vertically opposite to B
E	137°	Vertically opposite to G
F	43°	Corresponding to B
G	137°	Alternate to A
H	43°	Vertically opposite to F

3. a) i. Obtuse ii. Right iii. Reflex
   iv. Acute b) i. 70° ii. 110° iii. 110°
   iv. 70° v. 110°
4. $w = 35°$, $x = 70°$, $y = 75°$, $z = 75°$
5. $A = 25°$, $B = 140°$, $C = 40°$
6. $A = 23.5°$, $B = 83.5°$, $C = 23.5°$
7. $F = 45°$, Alternate; $G = 25°$, Alternate; $H = 70°$, Exterior angle is equal to the sum of the interior opposite angles
8. $K = 100°$, Vertically opposite; $L = 30°$, Alternate
13. a) $X = 40°$, Vertically opposite
    b) $Y = 70°$ c) $Z = 110°$
14. a) $X = 80°$ as opposite angles in a cyclic quadrilateral add to 180°
    b) $Y = 90°$

Exam-Style Questions
Question 1
$x = 10°$, $y = 30°$

Question 2
a) $a = 10°$ and $b = 30°$ b) i. 90°
ii. 75° d) $X = 30°$, $Y = 40°$ and $Z = 70°$

Question 3
a) 105° b) 36·6 m c) 73·2 m
d) €4557·15

Chapter 12: Applied Measure
1. a) Area = 6 cm², Perimeter = 12 cm
   b) Area = 60 mm², Perimeter = 38 mm c) Area = 25π cm², Perimeter = 10π cm
2. $52a^2$
3. a) 7 cm b) Radius = 6 cm, Circumference = 37·70 cm
   c) Length = 70 mm, Width = 80 mm, Area = 5600 mm²
4. a) $|AD| = 20$ cm b) 480 cm²
5. a) Area of the floor = 80 m², Area of the tiles = 0·0625 m², Total number of tiles needed = 1280
   b) €9052·80
6. a) Volume = 216 cm³
   b) 1.82×10^{25} atoms
7. a) 152 cm² b) 96 cm³
8. a) 1125π cm³ b) 942·48 cm²
   c) 1295·91 cm²
9. a) 4 cm b) 201·06 cm²
10. 770 cm²
11. 300 bowls; yes, the chef will have enough for all the guests
12. a) 0.33π cm³ b) 4·71 m²
13. 13 cm
14. a) 36π cm³ b) 4·4 cm

Exam-Style Questions
Question 1
a) Area of rectangle = 311·7691 m², Area of patio = 28·2743 m², Area of

ANSWERS 275

water feature = 3·14159 m², Area of triangles = 36√3, Area of grass = 218 m²   **b)** €15 750

Question 2
a) 407·15 cm³   **b) i.** Height = 20 cm, Width = 21·6   **ii.** 12 441·6 cm³
c) 78·5%   **d)** Height = 2 cans = 20 cm, Length = 2 cans = 14·4 cm, Width = 6 cans = 43·2 cm

Question 3
d) $\sqrt{7}$ **e)** 83 cm²   **f)** 21 cm³

Chapter 13: Statistics 1

1.
   - Primary data: the data you collect yourself
   - Secondary data: data that has already been collected and made available from an external source
   - Categorical data: non-numerical data
   - Nominal data: data that has no order or ranking e.g. colour, favourite musician
   - Ordinal data: can be sorted according to a category and have an order or ranking, e.g. exam grades, soccer leagues
   - Numerical data: numbers
   - Discrete data: data which can only have certain values, e.g. number of cars in a car park, number of people in a room
   - Continuous data: data which can take any numerical value, e.g. height, weight

3. Keep questions brief, use simple language, be clear where answers can be recorded, do not have questions leading in any way.
4. Large sample size, keep questions short, use simple language, choose a random sample
5. **b)** Discrete numerical data
   c) Incorrect, only asked 6-year-old girls. They did not include any boys or children aged 5, 7, 8 or 9.
   d) $\frac{1}{5}$ **e)** 33%
6. **a)**

Red	Blue	Pink	Purple	Orange
6	4	8	3	4

   c) Discrete numerical data because the data is whole numbers
   d) e.g. What is your favourite colour?
7. Physics = 90°, Applied Maths = 50°, Agricultural Science = 40°
8. **a)** Marbella = 40°, Santa Ponsa = 90°, Ibiza = 60°, Magaluf = 120°, Ayia Napa = 50°   **b) i.** Magaluf = 38 cm²   **ii.** Santa Ponsa = 28 cm²
9. **a)** 10°   **b)** C = 60 students, D = 25 students, E = 5 students   **c)** 180
   d) Discrete numerical

10. **a)** An outlier is a value in a data set that is very different from the other values; Outlier: 25   **b)** Mean
    c) Yes, as the range is the difference between the smallest and the largest value, either of which will be the outlier in a data set.
11. **a)** 22·625   **b)** 25   **c)** 24
    d) Mean, because it includes all values in the data set.   **e)** 17
    f) It tells us the difference between the largest and the smallest value
12. 84

Exam-Style Questions

Question 1
a) −1, 2, 2, 3, 14   **b)** 10   **c)** 3, 3, 9

Question 2
a) WW, DW, LW, WD, DD, LD, WL, DL, LL   **b)** $\frac{5}{9}$ **c)** 343   **d)** 1·9   **e)** 0 goals = 65°, 1 goal = 131°, 2 goals = 65°, 3 goals = 65°, 7 goals = 33°

Question 3
a) Ed Sheeran = 40 students, Dave = 20 students, Drake = 10 students, Miley Cyrus = 5 students   **c) i.** Discrete   **ii.** Discrete data is whole numbers
d) 65%   **e)** 45

Chapter 14: Statistics 2

1. **b)** 24   **c)** 36·17 minutes   **d)** 36–38 minutes   **e)** 34–36 minutes
   f) 83·33%   **g)** 11 people
   h) 4 people   **i)** Advantage: used to represent discrete or continuous data; Disadvantage: can be difficult to compare with other histograms, especially if different scales are used.
2. **b)** 40·1   **c)** 41·5   **d)** 50% of the workers are younger than 41·5 years and 50% are older.   **e)** 34   **f)** 47
3. **b) i.** Mean = 76·8, Mode = 70, Median = 73   **ii.** 47%   **iii.** Mean = 80·73, Mode = None, Median = 78   **iv.** 50% of students got less than 78 marks, 50% of students got more than 78 marks   **v.** Incorrect, we do not know who achieved each individual mark.

Exam-Style Questions

Question 1
a) 11·1 kg   **b)** 8·9 kg   **d)** Median (girls) = 8·7 kg, Median (boys) = 9·2 kg
e) No, the boys' mean is 8·9 kg and median is 9·2 kg whereas the girls' mean is 8·5 kg and median is 8·7 kg
f) £178   **g)** 516 km/hr

Question 2
a) $A = (2, 30)$, $B = (10, 6)$
b) $8\sqrt{10}$ units   **c)** 48·1 km
d)

0–10	10–20	20–30	30–40	40–50	50–60
3	4	8	7	17	11

f) 37·8 minutes   **g)** 40–50 interval
h) €41·66

Question 3
a)

0–2	2–4	4–6	6–8	8–10	10–12
11	31	18	13	11	3

12–14	14–16	16–18	18–20	20–22
1	1	6	1	4

b) 2–4 hours   **c)** 6·5 hours   **d) i.** $\frac{31}{100}$
ii. $\frac{3}{50}$ **iii.** 11%   **e)** Only asked first year boys, May have spent more time than usual on social media over the midterm break

Chapter 15: Probability

1. **a)** 12   **b)** 1H, 2H, 3H, 4H, 5H, 6H, 1T, 2T, 3T, 4T, 5T, 6T
2. 50: A0, A1, A2, A3, A4, A5, A6, A7, A8, A9, E0, E1, E2, E3, E4, E5, E6, E7, E8, E9, I0, I1, I2, I3, I4, I5, I6, I7, I8, I9, O0, O1, O2, O3, O4, O5, O6, O7, O8, O9, U0, U1, U2, U3, U4, U5, U6, U7, U8, U9
3. 16: YRH, YRF, YLH, YLF, BRH, BRF, BLH, BLF, GRH, GRF, GLH, GLF, RRH, RRF, RLH, RLF
4. **a)** A. Equally likely, B. Unlikely, C. Certain, D. Impossible, E. Likely
   b) D, B, A, E, C
5. **a)** 37·5%   **b)** Bigger sample space
6. Seán: $\frac{1}{4}$, 0·25, 25%   Megan: $\frac{2}{5}$, 0·4, 40%   Konrad: $\frac{3}{5}$, 0·6, 60%
7. **a)**

1	2	3	4	5	6
0·17	0·13	0·33	0·17	0·17	0·03

b) No, if it was fair all the relative frequencies would be equal.

8. **a)** $\frac{1}{2}$ **b)** $\frac{3}{13}$ **c)** $\frac{4}{13}$ **d)** $\frac{3}{13}$ **e)** $\frac{1}{4}$
   f) $\frac{3}{4}$
9. 0·2
10. 70%
11. **a)** $\frac{11}{24}$ **b)** $\frac{1}{6}$ **c)** $\frac{7}{24}$ **d)** $\frac{9}{13}$
12. **a)** 390   **b)** 195   **c)** 180   **d)** 15
    e) 60
13. **a)**

	2	4	6	8
3	5	7	9	11
5	7	9	11	13
7	9	11	13	15
9	11	13	15	17

b) i. $\frac{1}{4}$ **ii.** 0   **iii.** $\frac{5}{16}$ **iv.** $\frac{3}{8}$ **v.** $\frac{3}{16}$
14. **a)** $\frac{1}{8}$ **b)** $\frac{3}{8}$ **c)** $\frac{1}{2}$ **d)** $\frac{3}{8}$
15. **b) i.** $\frac{1}{6}$ **ii.** $\frac{7}{30}$ **iii.** $\frac{7}{30}$ **iv.** $\frac{11}{30}$ **v.** $\frac{23}{30}$
    vi. $\frac{3}{5}$
16. **b) i.** $\frac{41}{100}$ **ii.** $\frac{51}{100}$ **iii.** $\frac{3}{100}$ **iv.** $\frac{9}{100}$

v. $\frac{1}{20}$ vi. $\frac{1}{25}$ vii. $\frac{1}{5}$ viii. $\frac{14}{25}$

Exam-Style Questions

Question 1
b) 5   c) i. $\frac{9}{35}$ ii. $\frac{12}{35}$ iii. $\frac{1}{7}$

Question 2
b) 43   c) 40–60 interval   d) i. $\frac{1}{5}$
ii. $\frac{1}{6}$ e) $\frac{1}{4}$

Question 3
a) 0·6   b) 0·13   c) $A(2, 2)$ $B(9, 2)$
$C(9, 7)$ d) $|BA| = 7$, $|BC| = 5$,
therefore C is closest   e) $|AC| = \sqrt{74}$,
$|AC|^2 = |BA|^2 + |BC|^2$, therefore ABC is a
right-angled triangle   f) 35·54°

Chapter 16: Geometry 2

6. a) i. 120°   ii. 60°   iii. 60°
7. $\frac{6}{9} = \frac{10}{15} = \frac{8}{12}$ The triangles are similar
as the sides are in proportion
8. 21
9. b) i. 9 units   ii. 20 units   iii. 5 units
10. b) $\frac{|BR|}{|RD|} = \frac{|AB|}{|DC|} = \frac{|AR|}{|RC|}$ c) 5 cm
    d) 7 cm
11. 6 m
12. 295 m

Exam-Style Questions

Question 1
b) $|AB| = \sqrt{41}$ units $|BC| = \sqrt{41}$ units
c) It's an isosceles triangle as there are
2 equal sides   f) 51°

Question 2
a) i. 13 km   ii. 67·4°   iii. 22·6°
b) i. 12 km   ii. 5 km
c) Both courses are 30 km   d) 9 m/sec

Question 3
a) $x = 10°$, $y = 40°$
b) i. $|\angle BAC| = 50°$; $x + y$, $10 + 40 = 50°$
ii. $|\angle ABC| = 60°$; Given
iii. $|\angle ACB| = 70°$; Vertically opposite
iv. $|\angle DCE| = 70°$; Given
v. $|\angle CDE| = 60°$; Alternate to $|\angle ABC|$
vi. $|\angle CED| = 50°$; Alternate to $|\angle BAC|$
c) They are similar
d) $a = 4$ cm, $b = 9$ cm

Chapter 17: Geometry 3

15. c) Yes, they are congruent
17. c) Yes, they are similar
20. B, D, A, C
23. a) 2   b) 0   c) 1

Exam-Style Questions

Question 1
a) $x = 10°$, $y = 15°$ b) $|\angle BAD| = 110°$,
$|\angle ACD| = 70°$ c) i. 20 cm   ii. 16 cm

Question 2
b) $\frac{-1}{2} \times 2 = -1$, therefore the lines
are perpendicular to each other
d) $|FD|^2 = |ED|^2 + |EF|^2$, therefore the
triangle DEF is right angled

Question 3
c) $\sqrt{39}$ d) $|\angle ACB|$ is bisected as
$|\angle ACD| = 51·32°$ and $|\angle BCD| = 51·32°$